CAMBRIDGE MONOGRAPHS ON
MATHEMATICAL PHYSICS

General editors: P.V. Landshoff, D.R. Nelson, D.W. Sciama, S. Weinberg

DIFFERENTIAL GEOMETRY, GAUGE THEORIES,
AND GRAVITY

DIFFERENTIAL GEOMETRY, GAUGE THEORIES, AND GRAVITY

M. GÖCKELER
T. SCHÜCKER

Institute for Theoretical Physics, University of Heidelberg

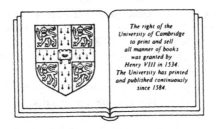

The right of the
University of Cambridge
to print and sell
all manner of books
was granted by
Henry VIII in 1534.
The University has printed
and published continuously
since 1584.

CAMBRIDGE UNIVERSITY PRESS

Cambridge

New York   Port Chester   Melbourne   Sydney

CAMBRIDGE UNIVERSITY PRESS
Cambridge, New York, Melbourne, Madrid, Cape Town, Singapore,
São Paulo, Delhi, Dubai, Tokyo, Mexico City

Cambridge University Press
The Edinburgh Building, Cambridge CB2 8RU, UK

Published in the United States of America by
Cambridge University Press, New York

www.cambridge.org
Information on this title: www.cambridge.org/9780521378215

First published 1987
First paperback edition 1989

A catalogue record for this publication is available from the British Library

Library of Congress Cataloguing in Publication Data

Gockeler, M.
Differential geometry, gauge theories, and gravity.
(Cambridge monographs on mathematical physics)
Bibliography: p.
Includes index.
1. Geometry, Differential. 2. Gauge theories
(Physics) 3. Gravity. 4. Mathematical physics.
I. Schücker, T. II. Title. III. Series.
QC20.7.D52G63 1987 516.3'6 87–11608

ISBN 978-0-521-32960-6 Hardback
ISBN 978-0-521-37821-5 Paperback

TO OUR PARENTS,
WHO TAUGHT US TO SEE BEAUTY
AND TO HONOUR TRUTH.

Irrtum verläßt uns nie, doch ziehet ein höher Bedürfnis
Immer den strebenden Geist leise zur Wahrheit hinan.

Error never leaves us, yet nevertheless a higher need constantly draws the
aspiring mind silently towards the truth.

GOETHE

Contents

Preface

This book is based on lecture notes of a three-semester course we taught at Heidelberg University during 1985–6. It is intended for graduate students in theoretical physics. The only prerequisites on the mathematical side are linear algebra and real analysis. The physical part, with the exception of the last chapter, is logically self-contained. However, since we give no motivations nor experimental applications, the reader should already be acquainted with the basics of Yang–Mills theories and general relativity.

The mathematical part is inspired by lectures André Haefliger taught at Geneva University. In chapters 1, 2, 3, and 6 we deal with differential forms and metric structures on $\mathbb{R}^n$. It is a mere rewriting of formulas well known to physicists from tensor analysis. This formalism, standard in mathematics, serves two purposes: It suppresses indices, which is an advantage in practical calculations; and it is coordinate-free, which allows straightforward generalization to topologically nontrivial spaces. Manifolds are introduced in chapter 7, Lie groups in chapter 8, and fibre bundles in chapter 9. Since fibre bundles are rather abstract mathematical objects and their relevance in physics is not (yet) established, we have organized the subsequent material in such a way that most of it can be understood without acquaintance with bundles. In particular, in chapter 11 covering spinors, we follow again the pattern: linear algebra, open subsets of $\mathbb{R}^n$, manifolds. The mathematics presented is essentially standard. Therefore we generally do not cite original work. For further details and proofs the reader is referred to the literature given in the bibliography, among other works.

Physical applications are presented in chapters 4, 5, 10, 12, and 13. The central theme is gauge invariance, motivated by today's belief that all known forces in Nature (gravitational, electromagnetic, and nuclear) are adequately described by gauge theories. For pedagogical reasons we first deal with Yang–Mills and Einstein–Cartan theories (chapters 4 and 5) on $\mathbb{R}^4$. However, written with differential forms they generalize easily to manifolds and bundles. We emphasize the fact that both theories are gauge theories. On the other hand, monopoles and instantons (chapter 10) are constructions involving nontrivial topology. Although it is unclear if they are necessary in physics, they provide nice examples for chapter 9 and explain the present use of bundle language in theoretical physics. Gauge anomalies are treated algebraically in chapter 12, first on $\mathbb{R}^4$ then for a general bundle. The last chapter, where we collect results of explicit anomaly calculations by Feynman graphs in $\mathbb{R}^n$, is disconnected from the main line, because we do not introduce quantum field theory.

We should like to express our gratitude to our teachers Kurt Meetz, André Haefliger, Henri Ruegg, and Raymond Stora. It is also a pleasure to acknowledge interaction with Hans Joos, Nikos Batakis, Flor Langouche, and Vaughan Jones.

M. Göckeler
T. Schücker

1987

1
Exterior algebra

In this chapter we repeat some parts of tensor algebra. For the applications we have in mind we only need a finite-dimensional subalgebra of the whole tensor algebra: the completely antisymmetric covariant tensors. These are most conveniently written in the language of forms.

1.1 Dual basis

Let V be a vector space over the real numbers of finite dimension $\dim V = n$. Its dual space V^* is by definition the set of forms, i.e. linear mappings from V into $\mathbb{R}$:

$$V^* := \{\varphi : V \to \mathbb{R}, \text{ linear}\}. \tag{1.1}$$

V^* is also a vector space of dimension n: If b_i, $i = 1, 2, \ldots, n$, is a basis of V, then the corresponding dual basis of V^*, β^i, $i = 1, 2, \ldots, n$, is defined by

$$\beta^i(b_j) := \delta^i{}_j. \tag{1.2}$$

Remember that a linear mapping is uniquely determined by its values on a basis.

Suppose we change to a new basis b'_i. The transformation from the old to the new basis is expressed by an invertible $n \times n$ matrix γ:

$$b'_i := \sum_{j=1}^{n} (\gamma^{-1})^j{}_i b_j, \quad (\gamma^{-1})^j{}_i \in \mathbb{R}. \tag{1.3}$$

In physics this is sometimes called 'passive transformation'. The set of all invertible $n \times n$ matrices forms a group denoted by GL_n. Its evolution during the different chapters should be followed closely because eventually it will become the gauge group of general relativity. If we denote β'^i the new dual basis, then the transformation is given by the inverse transposed matrix:

$$\beta'^i = \sum_{j=1}^{n} \gamma^i{}_j \beta^j. \tag{1.4}$$

1.2 Alternating forms

Let $\Lambda^p V$ be the vector space of p-linear alternating forms with real values:

$$\Lambda^p V := \left\{ \varphi : \underbrace{V \times V \times \cdots \times V}_{p \text{ factors}} \to \mathbb{R}, \text{ multilinear, alternating} \right\}. \quad (1.5)$$

Multilinear means linear in each of the p variables; alternating means that

$$\varphi(\ldots, v_i, \ldots, v_j, \ldots) = -\varphi(\ldots, v_j, \ldots, v_i, \ldots), \quad (1.6)$$

implying that a form evaluated on a linearly dependent set of vectors $v_1, \ldots, v_p$ is zero. $\Lambda^p V$ is the pth completely antisymmetric tensor power of V^*. In particular

$$\Lambda^1 V = V^*. \quad (1.7)$$

If p is bigger than n, the dimension of V, $\Lambda^p V$ is zero,

$$\Lambda^p V = 0 \quad \text{for} \quad p > n, \quad (1.8)$$

because then any p-tuple of vectors $v_1, \ldots, v_p$ is linearly dependent. By convention we set

$$\Lambda^0 V = \mathbb{R}. \quad (1.9)$$

A form $\varphi \in \Lambda^p V$ is said to be of degree p.

1.3 The wedge product

We define the wedge product (exterior product) of a p-form and a q-form to be a $(p+q)$-form:

$$\wedge : \Lambda^p V \times \Lambda^q V \to \Lambda^{p+q} V,$$

$$(\varphi, \psi) \mapsto \varphi \wedge \psi,$$

$$(\varphi \wedge \psi)(v_1, \ldots, v_{p+q}) :=$$

$$\frac{1}{p! q!} \sum_{\pi \in \mathscr{S}_{p+q}} \varphi(v_{\pi(1)}, \ldots, v_{\pi(p)}) \psi(v_{\pi(p+1)}, \ldots, v_{\pi(p+q)}) \operatorname{sig} \pi. \quad (1.10)$$

Here π runs over all permutations of $p+q$ objects and $\operatorname{sig} \pi$ denotes the sign of the permutation. For example, if φ and ψ are both 1-forms, we have

$$(\varphi \wedge \psi)(v_1, v_2) = \varphi(v_1)\psi(v_2) - \varphi(v_2)\psi(v_1) \quad (1.11)$$

and if φ is a 0-form, i.e. a real number, then $\varphi \wedge \psi = \varphi \psi$.

The following properties of the wedge product are derived by straight-forward calculations. It is

(a) bilinear, e.g.,

$$\begin{aligned}(\varphi_1 + \varphi_2) \wedge \psi &= \varphi_1 \wedge \psi + \varphi_2 \wedge \psi, \\ \varphi \wedge (a\psi) &= a(\varphi \wedge \psi), \quad a \in \mathbb{R};\end{aligned}\tag{1.12}$$

(b) associative,

$$(\varphi \wedge \psi) \wedge \chi = \varphi \wedge (\psi \wedge \chi);\tag{1.13}$$

(c) graded commutative,

$$\varphi \wedge \psi = (-1)^{pq} \psi \wedge \varphi, \quad \varphi \in \Lambda^p V, \quad \psi \in \Lambda^q V.\tag{1.14}$$

In particular, $\varphi \wedge \varphi = 0$ whenever φ is odd. A direct sum of vector spaces

$$\Lambda V := \bigoplus_{p=0}^{n} \Lambda^p V\tag{1.15}$$

together with a product satisfying (a), (b), and (c) is called exterior algebra or Grassmann algebra.

A basis of $\Lambda^p V$ is given by

$$\beta^{i_1} \wedge \beta^{i_2} \wedge \cdots \wedge \beta^{i_p}, \quad 1 \leqslant i_1 < i_2 < \cdots < i_p \leqslant n,\tag{1.16}$$

and therefore

$$\dim \Lambda^p V = \binom{n}{p},\tag{1.17}$$

$$\dim \Lambda V = 2^n.\tag{1.18}$$

Any p-form $\varphi \in \Lambda^p V$ can be decomposed in this basis:

$$\varphi = \sum_{i_1 < i_2 < \cdots < i_p} \varphi_{i_1 \cdots i_p} \beta^{i_1} \wedge \cdots \wedge \beta^{i_p}, \quad \varphi_{i_1 \cdots i_p} \in \mathbb{R},\tag{1.19}$$

or

$$\varphi = \frac{1}{p!} \sum_{i_1, \ldots, i_p} \varphi_{i_1 \cdots i_p} \beta^{i_1} \wedge \cdots \wedge \beta^{i_p}\tag{1.20}$$

with the convention that now $\varphi_{i_1 \cdots i_p}$ is defined for all values of indices (not only in increasing order) and that it is a completely antisymmetric tensor. To justify the qualification tensor we must consider a change of basis (1.3) under which the components of φ transform as a covariant tensor of rank p:

$$\varphi'_{i_1 \cdots i_p} = \sum_{j_1, \ldots, j_p} \varphi_{j_1 \cdots j_p} (\gamma^{-1})^{j_1}{}_{i_1} \cdots (\gamma^{-1})^{j_p}{}_{i_p}.\tag{1.21}$$

In component language the wedge product of a p-form φ and a q-form ψ corresponds to the antisymmetrized tensor product up to a factor $(p+q)!$:

$$\varphi \wedge \psi \leftrightarrow (p+q)!\,\varphi_{[i_1\cdots i_p}\psi_{j_1\cdots j_q]}. \tag{1.22}$$

The brackets indicate complete antisymmetrization.

1.4 The inner derivative

Our next subject is the inner derivative. Note that the word derivative refers to a purely algebraic property, the Leibniz rule.

For any vector v of the underlying vector space V there is one inner derivative i_v, which acts on forms lowering their degree by one unit:

$$i_v : \Lambda^p V \to \Lambda^{p-1} V$$

$$\varphi \mapsto i_v\varphi, \quad v \in V,$$

$$(i_v\varphi)(v_1,\ldots,v_{p-1}) := \varphi(v,v_1,\ldots,v_{p-1}). \tag{1.23}$$

In particular, if φ is a 1-form:

$$i_v\varphi = \varphi(v) \in \mathbb{R}. \tag{1.24}$$

In components this just corresponds to the contraction of the covariant tensor $\varphi_{i_1\cdots i_p}$ with the contravariant vector v^j:

$$i_v\varphi = \frac{1}{(p-1)!}\sum_{i_1,\ldots,i_{p-1}}\left(\sum_j v^j\varphi_{ji_1\cdots i_{p-1}}\right)\beta^{i_1}\wedge\cdots\wedge\beta^{i_{p-1}}, \quad v=\sum_j v^j b_j. \tag{1.25}$$

Immediate properties of the inner derivative are:

(a) i_v is a linear mapping;
(b) i_v is linear in v,

$$i_{v+w} = i_v + i_w, \tag{1.26}$$

$$i_{av} = a i_v, \quad a \in \mathbb{R}; \tag{1.27}$$

(c) (graded) Leibniz rule,

$$i_v(\varphi \wedge \psi) = (i_v\varphi)\wedge\psi + (-1)^p\varphi\wedge i_v\psi, \quad \varphi\in\Lambda^p V; \tag{1.28}$$

(d) $i_v i_w + i_w i_v = 0$, \hfill (1.29)

in particular,

$$i_v^2 = 0. \tag{1.30}$$

Equation (1.28) just states that i_v is a derivation of ΛV (see chapter 6).

1.5 Vector space-valued forms

The purpose of this section is to generalize real-valued forms to forms with values in a real vector space W. In later applications W will be the complex numbers, considered as two-dimensional real vector space, a Lie algebra (e.g. curvature is a Lie algebra valued 2-form) or a vector space carrying a representation of a Lie algebra. In supersymmetric theories, which we shall not discuss, forms with values in an abstract (infinite-dimensional) Grassmann algebra are considered.

$\Lambda^p(V, W)$ is the space of p-linear alternating forms with values in W:

$$\Lambda^p(V, W) := \left\{ \varphi : \underbrace{V \times \cdots \times V}_{p \text{ factors}} \to W, \text{ multilinear, alternating} \right\};$$

e.g. $\Lambda^1(V, W)$ is the set of linear mappings from V to W. Often it is useful to consider an element of $\Lambda^p(V, W)$ as a vector in W whose components are not just real numbers but real-valued p-forms. To this end let w_a be a basis of W. Any $\varphi \in \Lambda^p(V, W)$ can be decomposed:

$$\varphi = \sum_a \varphi^a w_a, \quad \varphi^a \in \Lambda^p V. \tag{1.31}$$

Note that in this general situation the wedge product is not defined. For instance in the product of two 1-forms $(\varphi \wedge \psi)(v_1, v_2) = \varphi(v_1)\psi(v_2) - \cdots$ we need to multiply the values $\varphi(v_1)$ and $\psi(v_2)$ which are now vectors in W. Only if W has a product, like the complex numbers, a Lie algebra or a Grassmann algebra, does the wedge product make sense. The multiplication of complex numbers being commutative, the wedge product for complex-valued forms is defined by (1.10) as in the real case.

Let us discuss in some detail the situation where W is a Lie algebra which we denote $\mathfrak{g}$. In this case we have to replace the products on the rhs of (1.10) by commutators. On the lhs we drop the wedge symbol and write $[\varphi, \psi]$. Alternatively, if T_a is a basis of $\mathfrak{g}$, we expand $\varphi \in \Lambda^p(V, \mathfrak{g})$ and $\psi \in \Lambda^q(V, \mathfrak{g})$,

$$\varphi = \sum_a \varphi^a T_a, \quad \psi = \sum_b \psi^b T_b,$$

$$\varphi^a \in \Lambda^p V, \quad \psi^b \in \Lambda^q V, \tag{1.32}$$

and get

$$[\varphi, \psi] = \sum_{a,b} \varphi^a \wedge \psi^b [T_a, T_b]. \tag{1.33}$$

From the antisymmetry of the commutator the graded commutativity gets an additional minus sign

$$[\varphi, \psi] = -(-1)^{pq}[\psi, \varphi]. \tag{1.34}$$

In particular, $[\varphi, \varphi]$ can be nonzero if φ is odd. The Jacobi identity becomes:

$$[[\varphi, \psi], \chi] + (-1)^{r(p+q)}[[\chi, \varphi], \psi] + (-1)^{p(r+q)}[[\psi, \chi], \varphi] = 0, \tag{1.35}$$

where

$$\varphi \in \Lambda^p(V, \mathfrak{g}), \quad \psi \in \Lambda^q(V, \mathfrak{g}), \quad \chi \in \Lambda^r(V, \mathfrak{g}).$$

1.6 Pull back

Previously we have defined 'passive transformations'. For completeness we now introduce 'active transformations'. As their formulas look very much alike the two are easily mixed up. We stress that as mathematical objects they are quite different.

Let $F : W \to V$ be a linear mapping between two vector spaces (W has nothing to do with the previous one). It induces a linear mapping

$$F^* : \Lambda^p V \to \Lambda^p W$$

$$\varphi \mapsto F^* \varphi$$

by

$$(F^*\varphi)(w_1, \ldots, w_p) := \varphi(Fw_1, \ldots, Fw_p). \tag{1.36}$$

F^* is called pull back.

Note that F is not necessarily invertible, and that the arrow in F^* now points in the opposite direction (from V to W). To write F^* in components we choose a basis b_i, $i = 1, 2, \ldots, n$ of V and a basis a_j, $j = 1, 2, \ldots, m$, in W. The components of F with respect to these two sets of basis vectors form the $n \times m$ matrix F:

$$F(a_j) = \sum_{i=1}^{n} F^i{}_j b_i. \tag{1.37}$$

The components of F^* with respect to the dual bases β^i and α^j are given by the transposition of F:

$$F^*(\beta^i) = \sum_{j=1}^{m} F^i{}_j \alpha^j. \tag{1.38}$$

We get the action of F^* on arbitrary forms by using the following properties:

(a) F^* is linear;

(b) $F^*(\varphi \wedge \psi) = (F^*\varphi) \wedge (F^*\psi)$; \hfill (1.39)

(c) $(F_1 \circ F_2)^* = F_2^* \circ F_1^*$; \hfill (1.40)

(a) and (b) state that F^* is a homomorphism of exterior algebras.

1.7 Orientation

This last section deals with the orientation of the vector space V, which we shall need in the next chapter to integrate forms. Our starting point is the remark that $\Lambda^n V$ is one-dimensional:

$$\dim \Lambda^n V = 1. \tag{1.41}$$

Indeed any (real-valued) n-form ω can be written as

$$\omega = k\beta^1 \wedge \cdots \wedge \beta^n$$

$$= \frac{k}{n!} \sum_{i_1,\ldots,i_n} \varepsilon_{i_1\cdots i_n} \beta^{i_1} \wedge \cdots \wedge \beta^{i_n} \tag{1.42}$$

where k is a real number and $\varepsilon_{i_1\cdots i_n}$ is the completely antisymmetric ε-symbol with

$$\varepsilon_{12\cdots n} = 1. \tag{1.43}$$

Note that k depends on the particular choice of the basis $\beta^1, \ldots, \beta^n$. Under a change of basis (1.4) it transforms as a density:

$$k' = k \det \gamma^{-1}. \tag{1.44}$$

By definition, an orientation of V is the choice of one of the two parts of $\Lambda^n V - \{0\}$, i.e. some n-form ω different from zero modulo a positive factor. Of course, any choice of basis $b_1, \ldots, b_n$ defines an orientation

$$\omega = \beta^1 \wedge \cdots \wedge \beta^n, \tag{1.45}$$

but any other basis b'_i such that its transformation matrix γ has positive determinant defines the same orientation. Conversely, given an orientation ω, an ordered basis $b_1, b_2, \ldots, b_n$ is called oriented if

$$\omega(b_1, b_2, \ldots, b_n) > 0. \tag{1.46}$$

Finally we mention the orientation induced by an active transformation. If $F : W \to V$ is an isomorphism (linear and bijective) and ω is an orientation of V, then $F^*\omega$ is by definition the induced orientation of W. In components:

$$F^*(\beta^1 \wedge \cdots \wedge \beta^n) = \det(F^i{}_j)\alpha^1 \wedge \cdots \wedge \alpha^n. \tag{1.47}$$

Problems

1.1 Verify the transformation law (1.4) of the dual basis.
1.2 Prove the Leibniz rule (1.28) for inner derivatives.
1.3 Derive the transformation property (1.44) of a density.

2
Differential forms on open subsets of $\mathbb{R}^n$

In the first chapter we have defined alternating p-forms φ on a vector space V. Now we are going to let these forms depend on a further argument x, which varies in an open subset of $\mathbb{R}^n$. Moreover, the resulting forms φ_x will not be applied to elements of a fixed vector space. Instead we have a different vector space for each x, the so-called tangent space at the point x. In this way we arrive at the notion of differential forms on open subsets of $\mathbb{R}^n$. The straightforward generalization to differential forms on arbitrary differentiable manifolds will be given in chapter 7.

2.1 Tangent vectors

Let $\mathscr{U}$ be an open subset of $\mathbb{R}^n$. The subset $\mathscr{U}$ may be described by Cartesian coordinates or any other coordinate system; i.e. we consider $x \in \mathscr{U}$ as a function of its coordinates (or, vice versa, the coordinates as functions of x), although usually we do not indicate this dependence explicitly. Which coordinates are used is either irrelevant or should be clear from the context.

The tangent space of $\mathscr{U}$ at the point $x \in \mathscr{U}$ is

$$T_x \mathscr{U} := \{x\} \times \mathbb{R}^n = \{(x, \xi) | \xi \in \mathbb{R}^n\}. \tag{2.1}$$

$T_x \mathscr{U}$ is an n-dimensional real vector space with linear structure defined by

$$a(x, \xi) + b(x, \eta) = (x, a\xi + b\eta), \quad a, b \in \mathbb{R}. \tag{2.2}$$

The elements of $T_x \mathscr{U}$ are called tangent vectors to $\mathscr{U}$ at x. Consider, for example, a parametrized curve in $\mathscr{U}$:

$$Q : [a, b] \to \mathscr{U}. \tag{2.3}$$

At each point $Q(\tau)$ of this curve we have the 'velocity' $dQ/d\tau \in \mathbb{R}^n$, which gives rise to the tangent vector

$$\dot{Q}(\tau) := \left(Q(\tau), \frac{dQ}{d\tau} \right) \in T_{Q(\tau)} \mathscr{U}. \tag{2.4}$$

Obviously, $dQ/d\tau$ can be an arbitrary element of $\mathbb{R}^n$, even if $Q(\tau)$ varies only in some 'small' $\mathcal{U}$. This shows that the $\mathbb{R}^n$ in the definition (2.1) of $T_x\mathcal{U}$ should not be mixed up with the $\mathbb{R}^n$ in which our subset $\mathcal{U}$ lies.

Tangent vectors always refer to some $x \in \mathcal{U}$, where they are located. Therefore we write them as pairs (x, ξ) with the point x as first entry. Note that (x, ξ) is not at all related to (y, ξ) if $x \neq y$: $(x, \xi) \in T_x\mathcal{U}$, $(y, \xi) \in T_y\mathcal{U}$, and $T_x\mathcal{U}$, $T_y\mathcal{U}$ are completely independent vector spaces. At each $x \in \mathcal{U}$ we have a separate copy of $\mathbb{R}^n$.

We can, however, consider the (disjoint) union of all these vector spaces, which is called the tangent bundle $T\mathcal{U}$ of $\mathcal{U}$:

$$T\mathcal{U} := \bigcup_{x \in \mathcal{U}} T_x\mathcal{U} = \{(x, \xi) \mid x \in \mathcal{U}, \xi \in \mathbb{R}^n\} = \mathcal{U} \times \mathbb{R}^n. \tag{2.5}$$

At this point, bundle is just a name. The notion of fibre bundles will be explained in chapter 9. Since $\mathcal{U} \times \mathbb{R}^n$ is an open subset of $\mathbb{R}^{2n}$, it is clear what is meant by continuous or differentiable mappings from or to $T\mathcal{U}$. We shall generally assume that all mappings we are dealing with are smooth, i.e. infinitely differentiable (C^∞). Of course, for a map Q as in (2.3) smoothness means that Q admits a smooth extension to an open interval containing $[a, b]$.

A smooth mapping v from $\mathcal{U}$ (or a subset of $\mathcal{U}$) into $T\mathcal{U}$ such that x is mapped onto a tangent vector at x is called a vector field:

$$x \mapsto (x, \xi(x)) = v(x) \in T_x\mathcal{U}. \tag{2.6}$$

As an example, one may think of the velocity field in a fluid.

A p-dimensional surface K in $\mathcal{U}$, given by a parameter representation, leads to vector fields which are tangent to K. To be more precise, let

$$Q: T \to \mathcal{U}, \quad T \subset \mathbb{R}^p,$$
$$\tau \mapsto Q(\tau), \tag{2.7}$$

be the parameter representation, which must be such that the $n \times p$ matrix

$$\left(\frac{\partial Q^i}{\partial \tau^j}\right)_{\substack{i=1,2,\dots,n \\ j=1,2,\dots,p}}$$

has rank p everywhere on T. We assume that Q can be smoothly extended to an open subset $\mathcal{V}$ of $\mathbb{R}^p$, which contains T. Now

$$Q(\tau) \mapsto \left(Q(\tau), \frac{\partial Q}{\partial \tau^j}\right), \quad j = 1, 2, \dots, p, \tag{2.8}$$

defines p vector fields on K, which are linearly independent at every point of K. They span a subspace of $T_{Q(\tau)}\mathcal{U}$, the tangent space of K.

2.2 Frames

Returning to $\mathcal{U}$ we call a set of n vector fields $b_i, i = 1,\dots,n$, which are linearly independent at every $x \in \mathcal{U}$, a frame (n-frame if we want to stress the dimension). Other names for this object are repère mobile, vielbein, comoving frame, and (for $n = 4$) tetrad. The $b_i(x)$ form a basis of $T_x\mathcal{U}$ which depends smoothly on $x \in \mathcal{U}$. If $x^1, x^2,\dots,x^n$ are coordinates on $\mathcal{U}$, they define an associated n-frame, denoted $\partial/\partial x^1, \partial/\partial x^2,\dots,\partial/\partial x^n$, by the formula

$$\frac{\partial}{\partial x^i}(x) = \frac{\partial}{\partial x^i}\bigg|_x := \left(x, \frac{\partial x}{\partial x^i}\right), \quad i = 1, 2,\dots,n. \tag{2.9}$$

(The symbol $|_x$ is just another way of indicating that a function, vector field,... is to be evaluated at the point x.) The notation $\partial/\partial x^i$ somehow anticipates the interpretation of tangent vectors as 'derivations' (see chapter 6). For the particular case of Cartesian coordinates $x^1, x^2,\dots,x^n$ one gets

$$\frac{\partial}{\partial x^1}\bigg|_x = (x,(1,0,\dots,0)),$$

$$\frac{\partial}{\partial x^2}\bigg|_x = (x,(0,1,0,\dots,0)), \tag{2.10}$$

$$\vdots$$

$$\frac{\partial}{\partial x^n}\bigg|_x = (x,(0,\dots,0,1)).$$

The vector field $\partial/\partial x^i$ is tangent to the coordinate line along which only x^i varies whereas the other coordinates are kept fixed. In fig. 2.1 these vector fields are shown for Cartesian coordinates x^1, x^2 in $\mathbb{R}^2$ and in fig. 2.2 for polar coordinates r, φ, where $x^1 = r\cos\varphi$, $x^2 = r\sin\varphi$. So every coordinate system on $\mathcal{U}$ determines a frame. However, given an arbitrary frame, there are in general no coordinates whose associated frame coincides with the given one.

Every vector field v may be decomposed with respect to a frame $b_1, b_2,\dots,b_n$:

$$v(x) = \sum_{i=1}^{n} v^i(x) b_i(x). \tag{2.11}$$

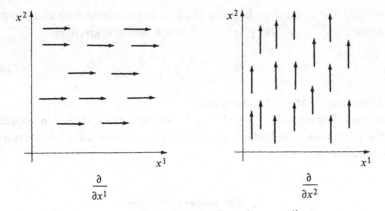

Fig. 2.1. Frame determined by Cartesian coordinates.

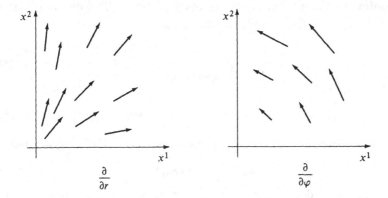

Fig. 2.2. Frame determined by polar coordinates.

The expansion coefficients v^i are smooth, real-valued functions. For the tangent vector (2.4) of a curve, the chain rule yields the representation

$$\dot{Q}(\tau) = \sum_{i=1}^{n} \frac{dQ^i}{d\tau} \frac{\partial}{\partial x^i} \bigg|_{Q(\tau)} \tag{2.12}$$

in terms of the frame $\partial/\partial x^1, \partial/\partial x^2, \ldots, \partial/\partial x^n$.

The connection between two frames $b_1, b_2, \ldots, b_n$ and $b'_1, b'_2, \ldots, b'_n$ is given by

$$b'_j(x) = \sum_{i=1}^{n} (\gamma^{-1}(x))^i{}_j \, b_i(x), \tag{2.13}$$

where the matrix $\gamma(x)$ is nondegenerate: $\gamma(x) \in GL_n$. (Attention: Do not confuse $\gamma^{-1}(x)$, the inverse of the matrix $\gamma(x)$, with an inverse mapping.) Note that a gauge group has appeared: γ is a group-valued function on $\mathcal{U}$. In particular, the connection between the frames associated with coordinate systems $x^1, x^2, \ldots, x^n$ and $y^1, y^2, \ldots, y^n$, respectively, is given by

$$\frac{\partial}{\partial y^j} = \sum_{i=1}^{n} \frac{\partial x^i}{\partial y^j} \frac{\partial}{\partial x^i}, \tag{2.14}$$

as follows easily from the chain rule.

In the following, $x^1, x^2, \ldots, x^n$ may be thought of as Cartesian coordinates, but unless stated otherwise, the formulas remain valid for arbitrary coordinates.

2.3 The tangent mapping

Consider the following situation. Let $\mathcal{U}$ be an open subset of $\mathbb{R}^n$ with coordinates $x^i, i = 1, 2, \ldots, n$, $\mathcal{V}$ an open subset of $\mathbb{R}^m$ with coordinates $y^j, j = 1, 2, \ldots, m$, and F a smooth mapping,

$$F: \mathcal{U} \to \mathcal{V}$$
$$x \mapsto F(x), \tag{2.15}$$

or in coordinates

$$(x^1, x^2, \ldots, x^n) \mapsto (F^1(x^1, \ldots, x^n), \ldots, F^m(x^1, \ldots, x^n)). \tag{2.16}$$

We want to define a corresponding linear mapping $T_x F$ of tangent spaces:

$$T_x F: T_x \mathcal{U} \to T_{F(x)} \mathcal{V}. \tag{2.17}$$

If $Q: [a, b] \to \mathcal{U}$ describes a curve in $\mathcal{U}$, $F \circ Q$ gives a curve in $\mathcal{V}$. At the parameter value τ we have the tangent vectors

$$\left(Q(\tau), \frac{dQ}{d\tau} \right) = \sum_{i=1}^{n} \frac{dQ^i}{d\tau} \frac{\partial}{\partial x^i} \bigg|_{Q(\tau)} \in T_{Q(\tau)} \mathcal{U} \tag{2.18}$$

and

$$\left((F \circ Q)(\tau), \frac{d}{d\tau}(F \circ Q) \right) = \sum_{j=1}^{m} \frac{d}{d\tau}(F \circ Q)^j \frac{\partial}{\partial y^j} \bigg|_{F(Q(\tau))} \in T_{F(Q(\tau))} \mathcal{V}. \tag{2.19}$$

We require that $T_{Q(\tau)} F$ maps (2.18) onto (2.19). Since

$$\frac{d}{d\tau}(F \circ Q)^j = \sum_{i=1}^{n} \frac{\partial F^j}{\partial x^i}(Q(\tau)) \frac{dQ^i}{d\tau} \tag{2.20}$$

we have to define

$$(T_x F) v = \sum_{j=1}^{m} \left(\sum_{i=1}^{n} v^i \frac{\partial}{\partial x^i} F^j \right) \frac{\partial}{\partial y^j}(F(x)) \tag{2.21}$$

for

$$v = \sum_{i=1}^{n} v^i \frac{\partial}{\partial x^i}(x) \in T_x \mathcal{U}.$$

$T_x F$ is called the tangent mapping. In the literature it is also denoted by F_*.
If we are given two maps F, G such that

$$\mathcal{U} \xrightarrow{F} \mathcal{V} \xrightarrow{G} \mathcal{W} \tag{2.22}$$

we have

$$T_x \mathcal{U} \xrightarrow{T_x F} T_{F(x)} \mathcal{V} \xrightarrow{T_{F(x)} G} T_{(G \circ F)(x)} \mathcal{W}, \tag{2.23}$$

and using the chain rule it is easy to check that

$$T_x(G \circ F) = T_{F(x)} G \circ T_x F. \tag{2.24}$$

2.4 Differential forms

We now come to one of the central notions of modern differential geometry.
A differential form of degree p $(p = 0, 1, \ldots, n)$, p-form for short, is a mapping
which maps every $x \in \mathcal{U}$ onto an alternating p-form φ_x on the tangent space
$T_x \mathcal{U}: \varphi_x \in \Lambda^p T_x \mathcal{U}$. So, applying φ_x to the tangent vectors $v_1, \ldots, v_p \in T_x \mathcal{U}$ we
get the real number $\varphi_x(v_1, \ldots, v_p)$. Since for a vector field v_i we have that
$v_i(x) \in T_x \mathcal{U}$, we can form $\varphi_x(v_1(x), \ldots, v_p(x))$. As x varies in $\mathcal{U}$, this expression
defines a real-valued function which is required to be smooth. The
generalization to forms with values in an arbitrary vector space is obvious.

By $\Lambda^p \mathcal{U}$ we denote the set of p-forms on $\mathcal{U}$ and hope that it will not be
confused with the set of p-forms on a vector space introduced in chapter 1.
Since according to (1.9) $\Lambda^0 T_x \mathcal{U} = \mathbb{R}$ we see that $\Lambda^0 \mathcal{U}$ is the set of all
(smooth) functions on $\mathcal{U}$. So a 0-form is simply a function. Defining a linear
structure on $\Lambda^p \mathcal{U}$ by

$$(a\varphi + b\psi)_x = a\varphi_x + b\psi_x \tag{2.25}$$

for $a, b \in \mathbb{R}$ and $\varphi, \psi \in \Lambda^p \mathcal{U}$, the set of p-forms on $\mathcal{U}$ becomes an infinite-
dimensional vector space over $\mathbb{R}$. Furthermore, we put

$$\Lambda \mathcal{U} := \bigoplus_{p=0}^{n} \Lambda^p \mathcal{U} \tag{2.26}$$

the direct sum of all the $\Lambda^p \mathcal{U}$.

The product of a function and a form, the wedge product of two forms
and the inner derivative of a form with respect to a vector field are defined
pointwise. They have the same algebraic properties as their analogues in
chapter 1 and are denoted by the same symbols.

We know already how to construct a basis for the space of p-forms over a vector space by means of a basis of this vector space and the wedge product (see (1.16)). Now we want to perform this construction at each $x \in \mathcal{U}$. Let $b_1, b_2, \ldots, b_n$ be a frame; i.e. $b_1(x), b_2(x), \ldots, b_n(x)$ is a basis of $T_x\mathcal{U}$. We denote the corresponding dual basis of $(T_x\mathcal{U})^* = \Lambda^1 T_x\mathcal{U}$, the cotangent space, by $\beta^1(x), \beta^2(x), \ldots, \beta^n(x)$. The set of 1-forms $\beta^1, \beta^2, \ldots, \beta^n$ is also called a frame. If $b'_1, b'_2, \ldots, b'_n$ is another frame, related to $b_1, b_2, \ldots, b_n$ by (2.13), the corresponding dual frames satisfy

$$\beta'^j(x) = \sum_{i=1}^{n} \gamma(x)^j{}_i \beta^i(x). \tag{2.27}$$

In the special case of the frame $\partial/\partial x^1, \partial/\partial x^2, \ldots, \partial/\partial x^n$ we write the dual frame as $dx^1, dx^2, \ldots, dx^n$. Omitting the argument x, as is usually done, we have

$$dx^i\left(\frac{\partial}{\partial x^j}\right) = \delta^i{}_j. \tag{2.28}$$

Here, $\delta^i{}_j$ means the constant function which takes the value $\delta^i{}_j$ for each $x \in \mathcal{U}$. A change of coordinates from $x^1, x^2, \ldots, x^n$ to $y^1, y^2, \ldots, y^n$ leads to a corresponding change of the associated frame in the cotangent space:

$$dy^i = \sum_{j=1}^{n} \frac{\partial y^i}{\partial x^j} dx^j. \tag{2.29}$$

By analogy with (1.20) we can represent an arbitrary $\varphi \in \Lambda^p \mathcal{U}$ as

$$\varphi = \frac{1}{p!} \sum_{i_1, \ldots, i_p = 1}^{n} \varphi_{i_1 \cdots i_p} \beta^{i_1} \wedge \beta^{i_2} \wedge \cdots \wedge \beta^{i_p}, \tag{2.30}$$

where the functions $\varphi_{i_1 \cdots i_p}$ are totally antisymmetric in their indices. In particular, an n-form may be written as

$$f\beta^1 \wedge \beta^2 \wedge \cdots \wedge \beta^n \tag{2.31}$$

with a function f. For later use we note that a 1-form

$$\varphi = \sum_{i=1}^{n} \varphi_i \beta^i \tag{2.32}$$

evaluated on a vector field

$$v = \sum_{i=1}^{n} v^i b_i \tag{2.33}$$

yields the function

$$\varphi(v) = \sum_{i=1}^{n} v^i \varphi_i = i_v \varphi. \tag{2.34}$$

2.5 The pullback of differential forms

If F is a mapping as in (2.15),

$$F: \mathcal{U} \to \mathcal{V}, \tag{2.35}$$

we can define the pullback operation

$$F^*: \Lambda^p \mathcal{V} \to \Lambda^p \mathcal{U}. \tag{2.36}$$

For $\varphi \in \Lambda^p \mathcal{V}$, $F^*\varphi$ is given by

$$(F^*\varphi)_x(v_1, \ldots, v_p) = \varphi_{F(x)}((T_xF)v_1, \ldots, (T_xF)v_p) \tag{2.37}$$

with $v_1, \ldots, v_p \in T_x \mathcal{U}$. Since $(T_xF)v_i \in T_{F(x)}\mathcal{V}$, the rhs is well defined and obviously an alternating multilinear form of the vs. Therefore (2.37) makes sense. For fixed x, the pullback of differential forms (2.37) corresponds to the pullback $(T_xF)^*$ of forms over a vector space (see (1.36)). In the special case $p = 0$, φ is a function and we have

$$(F^*\varphi)(x) = \varphi(F(x)); \tag{2.38}$$

i.e. $F^*\varphi = \varphi \circ F$.

The pullback of differential forms has the following properties, which are easily verified (compare (1.39), (1.40)):

(a) F^* is linear;
(b) $F^*(\varphi \wedge \psi) = (F^*\varphi) \wedge (F^*\psi);$ $\qquad$ (2.39)
(c) $(G \circ F)^* = F^* \circ G^*.$ $\qquad$ (2.40)

As an example we calculate $F^* \, dy^j$. (Remember that $y^1, y^2, \ldots, y^m$ are coordinates on $\mathcal{V}$.) Using the definition of F^* and T_xF one finds:

$$(F^* \, dy^j)\left(\frac{\partial}{\partial x^i}\right) = dy^j\left((T_xF)\frac{\partial}{\partial x^i}\right) = dy^j\left(\sum_{k=1}^{m} \frac{\partial F^k}{\partial x^i} \frac{\partial}{\partial y^k}\right) = \frac{\partial F^j}{\partial x^i} \tag{2.41}$$

and consequently

$$F^* \, dy^j = \sum_{i=1}^{n} \frac{\partial F^j}{\partial x^i} \, dx^i. \tag{2.42}$$

In physicists' language we can phrase this equation as: $F^* \, dy^j$ is the result of expressing the 'infinitesimal coordinate difference' dy^j in terms of the variables x^i where $y^j = F^j(x^1, x^2, \ldots, x^n)$.

In the special case $m = n$ (equal dimension of $\mathcal{U}$ and $\mathcal{V}$) the pullback of an

n-form on $\mathscr{V}$ is given by

$$F^*(f\,\mathrm{d}y^1 \wedge \cdots \wedge \mathrm{d}y^n) = f \circ F \det\left(\frac{\partial F^j}{\partial x^i}\right)\mathrm{d}x^1 \wedge \cdots \wedge \mathrm{d}x^n \qquad (2.43)$$

as is easily checked. Note the appearance of the Jacobian in (2.43), which indicates a connection between differential forms and integration.

2.6 The exterior derivative

The exterior derivative d, which we are now going to define, maps p-forms onto $(p+1)$-forms:

$$\mathrm{d}: \Lambda^p\mathscr{U} \to \Lambda^{p+1}\mathscr{U}. \qquad (2.44)$$

We start with the case $p = 0$, i.e., with the action of d on functions $f:\mathscr{U} \to \mathbb{R}$. The 1-form $\mathrm{d}f$ is given by

$$\mathrm{d}f := \sum_{i=1}^{n} \frac{\partial f}{\partial x^i}\mathrm{d}x^i. \qquad (2.45)$$

This definition does not depend on the choice of coordinates $x^1, x^2, \ldots, x^n$ as is easily shown by means of (2.29) and the chain rule.

The evaluation of $\mathrm{d}f$ on a tangent vector $v = \sum_{i=1}^{n}v^i\partial/\partial x^i \in T_x\mathscr{U}$ leads to the derivative of f in direction v:

$$(\mathrm{d}f)_x(v) = \sum_{i=1}^{n}\frac{\partial f}{\partial x^i}\mathrm{d}x^i\left(\sum_{j=1}^{n}v^j\frac{\partial}{\partial x^j}\right) = \sum_{i=1}^{n}v^i\frac{\partial f}{\partial x^i}. \qquad (2.46)$$

We shall often omit the argument x and write simply $\mathrm{d}f(v)$. If, in particular,

$$v = \dot{Q} = \sum_{i=1}^{n}\frac{\mathrm{d}Q^i}{\mathrm{d}\tau}\frac{\partial}{\partial x^i}$$

is a vector tangent to a curve in $\mathscr{U}$, we find

$$\mathrm{d}f(\dot{Q}) = \sum_{i=1}^{n}\frac{\mathrm{d}Q^i}{\mathrm{d}\tau}\frac{\partial f}{\partial x^i} = \frac{\mathrm{d}}{\mathrm{d}\tau}f(Q(\tau)). \qquad (2.47)$$

Applying d to the coordinate function x^i we get

$$\mathrm{d}(x^i) = \sum_{j=1}^{n}\frac{\partial x^i}{\partial x^j}\mathrm{d}x^j = \mathrm{d}x^i. \qquad (2.48)$$

So the notation $\mathrm{d}x^i$ for the basis of $(T_x\mathscr{U})^*$ dual to the coordinate basis $\partial/\partial x^j$ of $T_x\mathscr{U}$ is consistent with the above definition of the exterior derivative d.

One easily verifies the Leibniz rule

$$d(fg) = f\,dg + g\,df \tag{2.49}$$

for the exterior derivative of a product of two functions and the fact that d is linear:

$$d(af + bg) = a\,df + b\,dg \tag{2.50}$$

for $a, b \in \mathbb{R}$ and $f, g \in \Lambda^0 \mathcal{U}$.

The extension of d to an operator defined on p-forms with arbitrary p is accomplished by means of the following theorem.

THEOREM

There is one and only one map $d: \Lambda \mathcal{U} \to \Lambda \mathcal{U}$ with the properties:

(*a*) d is linear;
(*b*) $d(\Lambda^p \mathcal{U}) \subset \Lambda^{p+1} \mathcal{U}$;
(*c*) on functions $f \in \Lambda^0 \mathcal{U}$, d coincides with the operator introduced above;
(*d*) Leibniz rule:

$$d(\varphi \wedge \psi) = (d\varphi) \wedge \psi + (-1)^p \varphi \wedge d\psi, \quad \text{for} \quad \varphi \in \Lambda^p \mathcal{U}; \tag{2.51}$$

(*e*) $d^2 = 0$. \hfill (2.52)

Proof: Let $\varphi \in \Lambda^p \mathcal{U}$ be given by

$$\varphi = \frac{1}{p!} \sum_{i_1,\ldots,i_p = 1}^{n} \varphi_{i_1 \cdots i_p}\,dx^{i_1} \wedge \cdots \wedge dx^{i_p}. \tag{2.53}$$

With the help of (*a*), (*c*), (*d*) one gets

$$d\varphi = \frac{1}{p!} \sum_{i_1 \cdots i_p = 1}^{n} ((d\varphi_{i_1 \cdots i_p}) \wedge dx^{i_1} \wedge \cdots \wedge dx^{i_p}$$
$$+ \varphi_{i_1 \cdots i_p} d(dx^{i_1} \wedge \cdots \wedge dx^{i_p})), \tag{2.54}$$

where $d\varphi_{i_1 \cdots i_p}$ is the exterior derivative of the function $\varphi_{i_1 \cdots i_p}$. Using (*d*), (*e*) and the fact that $dx^i = d(x^i)$ as shown above, one finds

$$d(dx^{i_1} \wedge \cdots \wedge dx^{i_p}) = (ddx^{i_1}) \wedge dx^{i_2} \wedge \cdots \wedge dx^{i_p} - dx^{i_1} \wedge (ddx^{i_2})$$

$$\wedge dx^{i_3} \wedge \cdots \wedge dx^{i_p} + \cdots + (-1)^{p-1} dx^{i_1} \wedge \cdots \wedge dx^{i_{p-1}} \wedge ddx^{i_p} = 0 \tag{2.55}$$

and therefore

$$d\varphi = \frac{1}{p!} \sum_{i_1,\ldots,i_p = 1}^{n} (d\varphi_{i_1 \cdots i_p}) \wedge dx^{i_1} \wedge \cdots \wedge dx^{i_p}. \tag{2.56}$$

So, if an operator d with the required properties exists, it must be given by this formula, and uniqueness is proved.

Conversely, if we define d by (2.56), the conditions (a)–(e) are fulfilled as we shall demonstrate now. Statements (a)–(c) are obviously satisfied. To show (d) let

$$\varphi = \frac{1}{p!} \sum_{i_1,\ldots,i_p} \varphi_{i_1\cdots i_p} \, dx^{i_1} \wedge \cdots \wedge dx^{i_p} \in \Lambda^p \mathscr{U}, \tag{2.57}$$

$$\psi = \frac{1}{p!} \sum_{j_1,\ldots,j_q} \psi_{j_1\cdots j_q} \, dx^{j_1} \wedge \cdots \wedge dx^{j_q} \in \Lambda^q \mathscr{U}. \tag{2.58}$$

By means of (2.49) and (2.56) one calculates

$$d(\varphi \wedge \psi) = d \frac{1}{p!\,q!} \sum_{\substack{i_1,\ldots,i_p \\ j_1,\ldots,j_q}} \varphi_{i_1\cdots i_p} \psi_{j_1\cdots j_q} dx^{i_1} \wedge \cdots \wedge dx^{i_p} \wedge dx^{j_1} \wedge \cdots \wedge dx^{j_q}$$

$$= \frac{1}{p!\,q!} \sum_{\substack{i_1,\ldots,i_p \\ j_1,\ldots,j_q}} (d\varphi_{i_1\cdots i_p}) \psi_{j_1\cdots j_q} \wedge dx^{i_1} \wedge \cdots \wedge dx^{i_p} \wedge dx^{j_1} \wedge \cdots \wedge dx^{j_q}$$

$$+ \frac{1}{p!\,q!} \sum_{\substack{i_1,\ldots,i_p \\ j_1,\ldots,j_q}} \varphi_{i_1\cdots i_p} (d\psi_{j_1\cdots j_q}) \wedge dx^{i_1} \wedge \cdots \wedge dx^{i_p} \wedge dx^{j_1} \wedge \cdots \wedge dx^{j_q}$$

$$= \frac{1}{p!\,q!} \sum_{\substack{i_1,\ldots,i_p \\ j_1,\ldots,j_q}} (d\varphi_{i_1\cdots i_p}) \wedge dx^{i_1} \wedge \cdots \wedge dx^{i_p} \wedge \psi_{j_1\cdots j_q} dx^{j_1} \wedge \cdots \wedge dx^{j_q}$$

$$+ (-1)^p \frac{1}{p!\,q!} \sum_{\substack{i_1,\ldots,i_p \\ j_1,\ldots,j_q}} \varphi_{i_1\cdots i_p} dx^{i_1} \wedge \cdots \wedge dx^{i_p} \wedge (d\psi_{j_1\cdots j_q}) \wedge dx^{j_1} \wedge \cdots \wedge dx^{j_q}$$

$$= (d\varphi) \wedge \psi + (-1)^p \varphi \wedge d\psi. \tag{2.59}$$

For the proof of (e) we write (2.56) with the help of (2.45) as

$$d\varphi = \frac{1}{p!} \sum_{i,i_1,\ldots,i_p} \frac{\partial}{\partial x^i} \varphi_{i_1\cdots i_p} dx^i \wedge dx^{i_1} \wedge \cdots \wedge dx^{i_p} \tag{2.60}$$

and apply this formula once more:

$$dd\varphi = \frac{1}{p!} \sum_{i,j,i_1,\ldots,i_p} \frac{\partial^2}{\partial x^j \partial x^i} \varphi_{i_1\cdots i_p} dx^j \wedge dx^i \wedge dx^{i_1} \wedge \cdots \wedge dx^{i_p} = 0, \tag{2.61}$$

since the partial derivatives $\partial/\partial x^i, \partial/\partial x^j$ commute whereas $dx^i \wedge dx^j = -dx^j \wedge dx^i$. QED

Equation (2.60) may be rewritten as

$$d\varphi = \frac{1}{(p+1)!} \sum_{i_0,i_1,\ldots,i_p} \left(\frac{\partial}{\partial x^{i_0}} \varphi_{i_1\cdots i_p} - \frac{\partial}{\partial x^{i_1}} \varphi_{i_0 i_2 \cdots i_p} + \cdots \right.$$

$$+(-1)^p \frac{\partial}{\partial x^{i_p}} \varphi_{i_0 i_1 \cdots i_{p-1}} \Bigg) dx^{i_0} \wedge dx^{i_1} \wedge \cdots \wedge dx^{i_p}. \qquad (2.62)$$

This formula shows that $d\varphi$ is nothing but the antisymmetrized gradient of the coefficient functions $\varphi_{i_1 \cdots i_p}$.

According to the above method of construction, d is defined in a coordinate independent manner: $d\varphi$ is given by (2.60) for any coordinate system $x^1, x^2, \ldots, x^n$. Furthermore, d has a simple behaviour with respect to the pull back. Let $\mathcal{U} \subset \mathbb{R}^n$, $\mathcal{V} \subset \mathbb{R}^m$ be open sets and $F: \mathcal{U} \to \mathcal{V}$ a smooth mapping. Then we have

$$F^* d\varphi = dF^* \varphi \qquad (2.63)$$

for any differential form φ on $\mathcal{V}$. The proof is essentially an application of the chain rule. Note that d on the lhs of (2.63) is applied to forms on $\mathcal{V}$, whereas on the rhs it acts on elements of $\wedge \mathcal{U}$. So, to be more precise, one should write $d_{\mathcal{V}}$ and $d_{\mathcal{U}}$, respectively. However, since it is usually clear what is meant, we shall not make this distinction.

Let us now consider some special cases, where a connection with the operations of vector analysis may be established.

(a) The exterior derivative of a function f has components $\partial f / \partial x^i$ with respect to the frame $dx^1, dx^2, \ldots, dx^n$ (see (2.45)). If $x^1, x^2, \ldots, x^n$ are Cartesian coordinates, these are the components of $\mathrm{grad}\, f$.

(b) For a 2-form φ in $\mathbb{R}^3$, written as

$$\varphi = \tfrac{1}{2} \sum_{i,j,k=1}^{3} \varepsilon_{ijk} \varphi_i \, dx^j \wedge dx^k$$

$$= \varphi_1 \, dx^2 \wedge dx^3 + \varphi_2 \, dx^3 \wedge dx^1 + \varphi_3 \, dx^1 \wedge dx^2 \qquad (2.64)$$

we find

$$d\varphi = \sum_{i=1}^{3} \frac{\partial \varphi_i}{\partial x^i} dx^1 \wedge dx^2 \wedge dx^3. \qquad (2.65)$$

If x^1, x^2, x^3 are Cartesian coordinates and $\varphi_1, \varphi_2, \varphi_3$ are interpreted as the components of a vector field with respect to $\partial/\partial x^1$, $\partial/\partial x^2$, $\partial/\partial x^3$, we recognize in (2.65) the divergence of this vector field.

(c) Finally, let φ be a 1-form in $\mathbb{R}^3$:

$$\varphi = \sum_{i=1}^{3} \varphi_i \, dx^i. \qquad (2.66)$$

We get

$$d\varphi = \psi_1 \, dx^2 \wedge dx^3 + \psi_2 \, dx^3 \wedge dx^1 + \psi_3 \, dx^1 \wedge dx^2 \qquad (2.67)$$

with

$$\psi_1(x) = \frac{\partial \varphi_3}{\partial x^2} - \frac{\partial \varphi_2}{\partial x^3}, \quad \psi_2(x) = \frac{\partial \varphi_1}{\partial x^3} - \frac{\partial \varphi_3}{\partial x^1},$$

$$\psi_3(x) = \frac{\partial \varphi_2}{\partial x^1} - \frac{\partial \varphi_1}{\partial x^2}. \tag{2.68}$$

Again, if all components refer to a Cartesian basis, (2.68) represents the curl of a vector field.

With these interpretations in mind, it is easy to see that $d^2 \varphi = 0$ for a 0-form on $\mathbb{R}^3$ corresponds to curl grad $f = 0$, whereas the rule div curl $A = 0$ for a vector field A in $\mathbb{R}^3$ is contained in the statement $d^2 \varphi = 0$ with $\varphi \in \Lambda^1 \mathbb{R}^3$. A coordinate independent definition of grad, div, and curl requires the introduction of a metric and will be given in chapter 3.

2.7 The Poincaré lemma and de Rham cohomology

Forms φ which satisfy $d\varphi = 0$ are called closed. Note that any n-form on $\mathcal{U} \subset \mathbb{R}^n$ is closed, because the exterior derivative of an n-form has to be an $(n+1)$-form and the only $(n+1)$-form in n-dimensional space is identically zero. A p-form φ which can be written as the exterior derivative of a $(p-1)$-form χ,

$$\varphi = d\chi, \tag{2.69}$$

is called exact. Since $d^2 = 0$, every exact form is closed. However, a closed form is not necessarily exact, as the following (standard) example shows.

Let $\mathcal{U} = \mathbb{R}^2 - \{0\}$ be the Euclidean plane minus the origin and x, y Cartesian coordinates on $\mathbb{R}^2$. Then

$$\varphi = \frac{1}{x^2 + y^2}(x \, dy - y \, dx) \tag{2.70}$$

is a well-defined 1-form on $\mathcal{U}$. The exterior derivative of φ is easily calculated:

$$d\varphi = 0. \tag{2.71}$$

So φ is closed. Exactness would mean the existence of a smooth function f on $\mathcal{U}$ such that

$$\varphi = df. \tag{2.72}$$

Explicitly, (2.72) reads

$$\frac{\partial f}{\partial x} = -\frac{y}{x^2 + y^2}, \quad \frac{\partial f}{\partial y} = \frac{x}{x^2 + y^2}. \tag{2.73}$$

Up to an additive constant, the solution of these differential equations is given by

$$f(x, y) = \arctan(y/x), \tag{2.74}$$

i.e. the polar angle. However, f is not a continuous function on all of $\mathbb{R}^2 - \{0\}$. Therefore φ is not exact, although closed. Note that φ becomes exact if we consider it for example on the plane minus the negative x-axis, since in this domain f is a smooth function.

The following theorem gives a condition on $\mathscr{U}$, under which every closed form on $\mathscr{U}$ is exact.

THEOREM (Poincaré lemma)

If $\mathscr{U}$ is star-shaped, $\varphi \in \Lambda \mathscr{U}$ is closed if and only if it is exact.

An open subset $\mathscr{U}$ of $\mathbb{R}^n$ is called star-shaped if there is an $x_0 \in \mathscr{U}$ such that for any $x \in \mathscr{U}$ the straight line connecting x and x_0 is contained in $\mathscr{U}$. The Poincaré lemma is proved by explicit construction of a form whose exterior derivative equals the given closed φ. We restrict ourselves to the case of a 1-form $\varphi \in \Lambda^1 \mathscr{U}$ and assume that the x_0 in the definition of 'star-shaped' is the origin. In Cartesian coordinates x^i, φ may be written as

$$\varphi = \sum_{i=1}^{n} \varphi_i \, \mathrm{d}x^i. \tag{2.75}$$

Then the function

$$f(x) = \int_0^1 \sum_{i=1}^{n} \varphi_i(tx) x^i \, \mathrm{d}t \tag{2.76}$$

satisfies the equation

$$\mathrm{d}f = \varphi \tag{2.77}$$

as is easily checked using $\mathrm{d}\varphi = 0$. Note that the argument of the functions φ_i in the integral (2.76) runs along a straight line from 0 to x as t varies between 0 and 1. Since $\mathscr{U}$ is assumed to be star-shaped with $x_0 = 0$, these straight lines lie completely within $\mathscr{U}$ for any $x \in \mathscr{U}$, i.e. in the domain of definition of the φ_i. Note also that the above construction is used in physics to get a potential of a conservative force field. The proof of the Poincaré lemma for $\varphi \in \Lambda^p \mathscr{U}$ with $p > 1$ is similar.

The example (2.70) as well as the Poincaré lemma indicate that the existence of forms which are closed but not exact has something to do with the geometry of the domain $\mathcal{U}$ on which they are considered. This observation is the starting point of de Rham cohomology. Denote by $Z^p(\mathcal{U})$ the real vector space of closed p-forms on $\mathcal{U}$:

$$Z^p(\mathcal{U}) := \{\varphi \in \Lambda^p \mathcal{U} \mid \varphi \text{ closed}\}. \tag{2.78}$$

For $p > 0$, let $B^p(\mathcal{U})$ be the real vector space of exact p-forms on $\mathcal{U}$:

$$B^p(\mathcal{U}) := \{\varphi \in \Lambda^p \mathcal{U} \mid \varphi \text{ exact}\}. \tag{2.79}$$

Furthermore, put

$$B^0(\mathcal{U}) := \{0\}. \tag{2.80}$$

Since $B^p(\mathcal{U}) \subset Z^p(\mathcal{U})$, we can define the de Rham cohomology groups

$$H^p(\mathcal{U}) := Z^p(\mathcal{U})/B^p(\mathcal{U}), \tag{2.81}$$

which are not only abelian groups but also vector spaces. The dimension of $H^p(\mathcal{U})$, if finite, is called the pth Betti number of $\mathcal{U}$ and is denoted by $b^p(\mathcal{U})$. The alternating sum of the Betti numbers is the Euler characteristic

$$\chi(\mathcal{U}) := \sum_{p=0}^{n} (-1)^p b^p(\mathcal{U}).$$

2.8 Integration of n-forms

Now we come to the important topic of integration of differential forms. We shall define integration over oriented spaces only. So we have to start with a discussion of the concept of orientation. Recall that an orientation of an n-dimensional vector space corresponds to the choice of a non-vanishing n-form ω modulo a positive factor or, equivalently, to the choice of an ordered basis $b_1, b_2, \ldots, b_n$.

An open subset $\mathcal{U}$ of $\mathbb{R}^n$ is oriented by defining a coherent orientation on the tangent spaces $T_x \mathcal{U}$. This can be done by giving a nowhere vanishing n-form $\omega \in \Lambda^n \mathcal{U}$. Then we have for an arbitrary frame $b_1, b_2, \ldots, b_n$:

$$\omega_x(b_1(x), b_2(x), \ldots, b_n(x)) \neq 0 \quad \text{for all} \quad x \in \mathcal{U}. \tag{2.82}$$

We call the ordered frame $b_1, b_2, \ldots, b_n$ oriented if

$$\omega_x(b_1(x), b_2(x), \ldots, b_n(x)) > 0 \quad \text{for all} \quad x \in \mathcal{U}. \tag{2.83}$$

The notion of orientation is also applied to ordered coordinate systems $x^1, x^2, \ldots, x^n$ on $\mathcal{U}$. They are called oriented if the associated frames are oriented:

$$\omega_x\left(\frac{\partial}{\partial x^1}(x),\ldots,\frac{\partial}{\partial x^n}(x)\right)>0,\quad x\in\mathcal{U}. \tag{2.84}$$

Conversely, an orientation of $\mathcal{U}$ may be given by declaring a coordinate system $x^1,x^2,\ldots,x^n$ to be oriented. Then one can take as orienting n-form

$$\omega=dx^1\wedge dx^2\wedge\cdots\wedge dx^n. \tag{2.85}$$

If f is a function on $\mathcal{U}$ such that $f(x)>0$ for all $x\in\mathcal{U}$, then $f\omega$ and ω lead to the same orientation. So, strictly speaking, an orientation is an equivalence class of n-forms, where two n-forms are equivalent if they differ by an everywhere positive factor.

We first define the integral of an n-form over a sufficiently regular subset K of $\mathcal{U}$. In order to avoid lengthy technical discussions we do not specify the precise meaning of 'sufficiently regular' and simply say: K has to be such that the ordinary integrals, which will appear, are well defined.

Now let $x^1,x^2,\ldots,x^n$ be an oriented coordinate system on $\mathcal{U}$ and $\varphi\in\Lambda^n\mathcal{U}$. We can write

$$\varphi=f\,dx^1\wedge dx^2\wedge\cdots\wedge dx^n, \tag{2.86}$$

where f is a function on $\mathcal{U}$, given by

$$f(x)=\varphi_x\left(\frac{\partial}{\partial x^1}(x),\ldots,\frac{\partial}{\partial x^n}(x)\right). \tag{2.87}$$

The integral of the n-form φ over $K\subset\mathcal{U}$ is defined as

$$\int_K\varphi:=\int\int\cdots\int_K f(x)\,dx^1\,dx^2\cdots dx^n. \tag{2.88}$$

The expression on the rhs of this equation is to be interpreted as the multiple or volume integral of standard calculus, extended over the domain in the space of coordinates that corresponds to K. To simplify notation we have denoted this domain by K, too. Obviously, the integral of n-forms is linear:

$$\int_K(a\varphi+b\psi)=a\int_K\varphi+b\int_K\psi \tag{2.89}$$

for $a,b\in\mathbb{R}$ and $\varphi,\psi\in\Lambda^n\mathcal{U}$.

If one reverses the orientation, e.g. by taking $x^2,x^1,x^3,\ldots,x^n$ as an oriented coordinate system, the integral changes sign, since we now have to write

$$\varphi = (-f)\, dx^2 \wedge dx^1 \wedge dx^3 \wedge \cdots \wedge dx^n \qquad (2.90)$$

with the same function f as in (2.86). Therefore we get

$$\int_K \varphi = \int\int \cdots \int_K (-f(x))\, dx^2\, dx^1\, dx^3 \cdots dx^n$$

$$= -\int\int \cdots \int_K f(x)\, dx^1\, dx^2 \cdots dx^n. \qquad (2.91)$$

Definition (2.88) looks coordinate-dependent. But it is not, due to the transformation properties of the frame $dx^1, dx^2, \ldots, dx^n$. To see this let $y^1, y^2, \ldots, y^n$ be another oriented coordinate system on $\mathcal{U}$. Then we have (see (2.29)):

$$dy^1 \wedge dy^2 \wedge \cdots \wedge dy^n = \det\left(\frac{\partial y^i}{\partial x^j}\right) dx^1 \wedge dx^2 \wedge \cdots \wedge dx^n \qquad (2.92)$$

with $\det(\partial y^i/\partial x^j) > 0$, and φ is written as

$$\varphi = f(x(x^1, \ldots, x^n))\, dx^1 \wedge \cdots \wedge dx^n$$

$$= f(x(y^1, \ldots, y^n)) \det\left(\frac{\partial x^i}{\partial y^j}\right) dy^1 \wedge \cdots \wedge dy^n. \qquad (2.93)$$

Using the coordinates x^i we find

$$\int_K \varphi = \int\int \cdots \int_K f(x(x^1, \ldots, x^n))\, dx^1 \cdots dx^n, \qquad (2.94)$$

whereas the coordinates y^j lead to

$$\int_K \varphi = \int\int \cdots \int_K f(x(y^1, \ldots, y^n)) \det\left(\frac{\partial x^i}{\partial y^j}\right) dy^1 \cdots dy^n. \qquad (2.95)$$

That these two expressions for $\int_K \varphi$ are equal is the statement of the well-known rule of change of variables in multiple integrals. (Note that the Jacobian is positive!)

Next we want to study the behaviour of integrals under diffeomorphisms. So let $\mathcal{U}$ be an open subset of $\mathbb{R}^n$ with oriented coordinate system $x^1, x^2, \ldots, x^n$ and $\mathcal{V}$ an open subset of $\mathbb{R}^n$ with $y^1, y^2, \ldots, y^n$ as the oriented coordinate system. Consider a diffeomorphism

$$F: \mathcal{U} \to \mathcal{V}; \qquad (2.96)$$

i.e. F is bijective, and F as well as F^{-1} are smooth. Furthermore, we assume that F is orientation-preserving; i.e. $F^* dy^1 \wedge \cdots \wedge dy^n$ defines the same orientation on $\mathscr{U}$ as $dx^1 \wedge \cdots \wedge dx^n$. Since

$$F^* dy^1 \wedge \cdots \wedge dy^n = \det\left(\frac{\partial F^j}{\partial x^i}\right) dx^1 \wedge \cdots \wedge dx^n, \qquad (2.97)$$

this means:

$$\det\left(\frac{\partial F^j}{\partial x^i}\right) > 0. \qquad (2.98)$$

Under these assumptions we have for sufficiently regular $K \subset \mathscr{U}$ and $\varphi \in \Lambda^n \mathscr{V}$:

$$\int_K F^* \varphi = \int_{F(K)} \varphi. \qquad (2.99)$$

In order to prove this equation we write

$$\varphi = f \, dy^1 \wedge \cdots \wedge dy^n. \qquad (2.100)$$

Consequently

$$F^* \varphi = (f \circ F) \det\left(\frac{\partial F^j}{\partial x^i}\right) dx^1 \wedge \cdots \wedge dx^n, \qquad (2.101)$$

and we get

$$\int_K F^* \varphi = \int\int \cdots \int_K f(F(x)) \det\left(\frac{\partial F^j}{\partial x^i}\right) dx^1 \, dx^2 \cdots dx^n, \qquad (2.102)$$

$$\int_{F(K)} \varphi = \int\int \cdots \int_{F(K)} f(y) \, dy^1 \, dy^2 \cdots dy^n. \qquad (2.103)$$

Once more we recognize the rule of change of variables in multiple integrals. But this time we are dealing with an active transformation F: It maps a point x onto a (generally) different point $F(x)$. That is why we have written the domain of integration as $F(K)$ on the rhs of (2.99). In equations (2.94), (2.95), on the other hand, we considered the map

$$(x^1, \ldots, x^n) \mapsto (y^1(x^1, \ldots, x^n), \ldots, y^n(x^1, \ldots, x^n)) \qquad (2.104)$$

as a coordinate transformation (passive point of view): $x^1, \ldots, x^n$ and $y^1(x^1, \ldots, x^n), \ldots, y^n(x^1, \ldots, x^n)$ describe the same point. Consequently, we denoted the domain of integration by the same letter K in (2.94) and (2.95).

2.9 Integration of p-forms

Up to now we have only defined integrals of n-forms over n-dimensional domains in $\mathbb{R}^n$. We want to generalize this definition to include integrals of p-forms over p-dimensional oriented surfaces K in $\mathbb{R}^n$. By an oriented surface we mean a surface given by a smooth parameter representation $Q: T \to \mathcal{U}$ as in (2.7) such that the open set $\mathcal{V}$ containing T is an oriented open subset of $\mathbb{R}^p$. Furthermore, we have to assume that T is a sufficiently regular subset of the parameter space $\mathcal{V}$.

For a p-form φ on $\mathcal{U}$ we define

$$\int_K \varphi := \int_T Q^*\varphi. \tag{2.105}$$

On the rhs of this equation we have an integral of a p-form over a p-dimensional domain in $\mathbb{R}^p$ as defined in section 2.8. If $p = n$ and K is a sufficiently regular subset of $\mathcal{U}$, we can choose a parameter representation such that $\mathcal{V} = \mathcal{U}$, $T = K$ and $Q =$ inclusion map. Therefore (2.105) is compatible with our former definition (2.88). Since a change of the parameter representation is given by an orientation-preserving diffeomorphism of the parameter spaces, (2.99) shows that $\int_K \varphi$ as defined in (2.105) is independent of the chosen parameter representation.

It might be impossible to cover the whole surface, over which one wants to integrate, with one parameter representation. In such a case one cuts the surface into pieces which can be described by a single parameter representation and integrates over each of them separately. Of course, one has to choose one coherent orientation for the whole surface. So we must restrict ourselves to surfaces for which this is possible (orientable surfaces). The well-known Möbius strip, for example, cannot be oriented.

For illustration we consider two examples.

(a) $p = 1$. In this case, K is a curve with parameter representation

$$Q: [a, b] \to \mathcal{U},$$
$$\tau \mapsto Q(\tau). \tag{2.106}$$

The orientation is fixed by declaring the coordinate τ to form an oriented coordinate system. Let $x^1, x^2, \ldots, x^n$ be Cartesian coordinates on $\mathcal{U}$. Then $\varphi \in \Lambda^1 \mathcal{U}$ can be written as

$$\varphi = \sum_{i=1}^n \varphi_i \, \mathrm{d}x^i \tag{2.107}$$

and we have

$$Q^*\varphi = \sum_{i=1}^n (\varphi_i \circ Q) \frac{\mathrm{d}Q^i}{\mathrm{d}\tau} \, \mathrm{d}\tau. \tag{2.108}$$

So we find

$$\int_K \varphi = \int_a^b \sum_{i=1}^n \varphi_i(Q(\tau)) \frac{dQ^i}{d\tau} d\tau. \tag{2.109}$$

If $\varphi_1, \ldots, \varphi_n$ are interpreted as Cartesian components of a vector field on $\mathscr{U}$, (2.109) is equal to the standard line integral of a vector field.

(b) $p = 2$, $n = 3$. Now K is a two-dimensional surface in $\mathscr{U} \subset \mathbb{R}^3$. We choose Cartesian coordinates x^1, x^2, x^3 on $\mathscr{U}$ and a parameter representation

$$Q : T \to \mathscr{U}$$
$$(\tau^1, \tau^2) \mapsto (Q^1(\tau^1, \tau^2), Q^2(\tau^1, \tau^2), Q^3(\tau^1, \tau^2)), \tag{2.110}$$

where τ^1, τ^2 is an oriented coordinate system on $\mathscr{V}$, the open set containing T. Again, the 2-form φ on $\mathscr{U}$ is written as

$$\varphi = \frac{1}{2} \sum_{i,j,k=1}^3 \varepsilon_{ijk} \varphi_i \, dx^j \wedge dx^k. \tag{2.111}$$

Using the vector product in $\mathbb{R}^3$ we get

$$Q^* \varphi = \sum_{i=1}^3 (\varphi_i \circ Q) \left(\frac{\partial Q}{\partial \tau^1} \times \frac{\partial Q}{\partial \tau^2} \right)^i d\tau^1 \wedge d\tau^2 \tag{2.112}$$

and consequently

$$\int_K \varphi = \int_K (\varphi_1 \, dx^2 \wedge dx^3 + \varphi_2 \, dx^3 \wedge dx^1 + \varphi_3 \, dx^1 \wedge dx^2)$$
$$= \int_T \sum_{i=1}^3 \varphi_i(Q(\tau)) \left(\frac{\partial Q}{\partial \tau^1} \times \frac{\partial Q}{\partial \tau^2} \right)^i d\tau^1 \, d\tau^2. \tag{2.113}$$

We recognize the 'vectorial surface element'

$$\frac{\partial Q}{\partial \tau^1} \times \frac{\partial Q}{\partial \tau^2} d\tau^1 \, d\tau^2$$

and the surface integral of the vector field with Cartesian components $\varphi_1, \varphi_2, \varphi_3$.

2.10 Stokes' theorem

Our last topic in this chapter is the (generalized) Stokes theorem, which includes, as special cases, many of the classical integral theorems like the theorems of Gauss and Stokes. In order to formulate it, we have to take a closer look at the boundaries of our integration domains. In particular, we must provide them with an orientation.

Let $K \subset \mathscr{U}$ be a p-dimensional oriented surface described by the parameter representation

$$Q: T \to \mathscr{U}, \tag{2.114}$$

where $T \subset \mathscr{V}$ and $\mathscr{V}$ is an oriented open subset of $\mathbb{R}^p$. By ∂K we denote the boundary of K. It is a $(p-1)$-dimensional surface built out of pieces $(\partial K)_i$ which possess a parameter representation. Take $\tau^1, \dots, \tau^p$ as an oriented coordinate system on $\mathscr{V}$. Then the vectors

$$\left(Q(\tau), \frac{\partial Q}{\partial \tau^1} \right), \dots, \left(Q(\tau), \frac{\partial Q}{\partial \tau^p} \right) \tag{2.115}$$

span the tangent space of K at $Q(\tau)$. They form an oriented p-frame on K.

The orientation of K induces an orientation of ∂K in the following way. Let n be an outward normal vector field on ∂K; i.e. $n(x)$ is tangent to K at $x \in \partial K$ and directed to the exterior of K. Choose the parameters $\tilde{\tau}^1, \dots, \tilde{\tau}^{p-1}$ in the parameter representation

$$\tilde{Q}: \tilde{T} \to \mathscr{U}, \quad \tilde{T} \subset \mathbb{R}^{p-1} \tag{2.116}$$

of $(\partial K)_i$ such that

$$n(\tilde{Q}(\tilde{\tau})), \quad \left(\tilde{Q}(\tilde{\tau}), \frac{\partial \tilde{Q}}{\partial \tilde{\tau}^1} \right), \dots, \left(\tilde{Q}(\tilde{\tau}), \frac{\partial \tilde{Q}}{\partial \tilde{\tau}^{p-1}} \right) \tag{2.117}$$

is an oriented basis of the tangent space of K at $\tilde{Q}(\tilde{\tau})$. It can be proved that this procedure indeed leads to a well-defined orientation on ∂K. Fig. 2.3 illustrates our construction for $p = 2$, $n = 3$. In the following, ∂K is always assumed to be oriented in this way if K is oriented.

After these preparations we are ready to formulate Stokes' theorem:

$$\int_K \mathrm{d}\varphi = \int_{\partial K} \varphi, \tag{2.118}$$

where K is a p-dimensional oriented surface and $\varphi \in \Lambda^{p-1}\mathscr{U}$.

Let us consider some special cases.

(a) $p = n = 2$ with K a sufficiently regular subset of $\mathscr{U} = \mathbb{R}^2$ (see fig. 2.4). Let the orientation be given by the Cartesian coordinates x^1, x^2 on $\mathscr{U}$. For $\varphi = \varphi_1 \,\mathrm{d}x^1 + \varphi_2 \,\mathrm{d}x^2 \in \Lambda^1 \mathscr{U}$ we find

$$\mathrm{d}\varphi = \left(\frac{\partial \varphi_2}{\partial x^1} - \frac{\partial \varphi_1}{\partial x^2} \right) \mathrm{d}x^1 \wedge \mathrm{d}x^2. \tag{2.119}$$

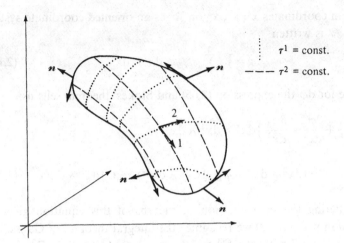

Fig. 2.3. Orientation of the boundary of a two-dimensional oriented surface.

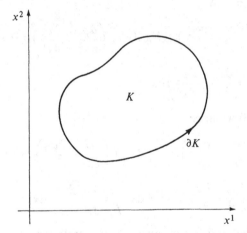

Fig. 2.4. Orientation of the boundary of $K \subset \mathbb{R}^2$.

Consequently,

$$\int_K d\varphi = \iint_K \left(\frac{\partial \varphi_2}{\partial x^1} - \frac{\partial \varphi_1}{\partial x^2} \right) dx^1 \, dx^2, \qquad (2.120)$$

and Stokes' theorem gives the well-known result (Green's theorem)

$$\iint_K \left(\frac{\partial \varphi_2}{\partial x^1} - \frac{\partial \varphi_1}{\partial x^2} \right) dx^1 \, dx^2 = \int_{\partial K} (\varphi_1(x) \, dx^1 + \varphi_2(x) \, dx^2). \qquad (2.121)$$

(b) $p = n = 3$ with K a sufficiently regular subset of $\mathcal{U} = \mathbb{R}^3$. Take the

Cartesian coordinates x^1, x^2, x^3 on $\mathbb{R}^3$ as an oriented coordinate system. If $\varphi \in \Lambda^2 \mathcal{U}$ is written

$$\varphi = \tfrac{1}{2} \sum_{i,j,k=1}^{3} \varepsilon_{ijk} \varphi_i \, dx^j \wedge dx^k, \tag{2.122}$$

we have for $d\varphi$ the expression (2.65) and Stokes' theorem tells us:

$$\int_K \left(\frac{\partial \varphi_1}{\partial x^1} + \frac{\partial \varphi_2}{\partial x^2} + \frac{\partial \varphi_3}{\partial x^3} \right) dx^1 \wedge dx^2 \wedge dx^3$$

$$= \int_{\partial K} (\varphi_1(x) \, dx^2 \wedge dx^3 + \varphi_2(x) \, dx^3 \wedge dx^1 + \varphi_3(x) \, dx^1 \wedge dx^2). \tag{2.123}$$

Remembering the interpretation of the rhs of this equation given in connection with (2.113) we recognize the integral theorem of Gauss.

(c) $p = 2$, $n = 3$ with $\mathcal{U} = \mathbb{R}^3$ and K an oriented surface in $\mathbb{R}^3$. Choose Cartesian coordinates x^1, x^2, x^3 in $\mathbb{R}^3$. According to (2.67), (2.68) we get for $\varphi = \sum_{i=1}^{3} \varphi_i \, dx^i \in \Lambda^1 \mathcal{U}$

$$d\varphi = \left(\frac{\partial \varphi_3}{\partial x^2} - \frac{\partial \varphi_2}{\partial x^3} \right) dx^2 \wedge dx^3 + \left(\frac{\partial \varphi_1}{\partial x^3} - \frac{\partial \varphi_3}{\partial x^1} \right) dx^3 \wedge dx^1$$

$$+ \left(\frac{\partial \varphi_2}{\partial x^1} - \frac{\partial \varphi_1}{\partial x^2} \right) dx^1 \wedge dx^2, \tag{2.124}$$

and from (2.118) we infer

$$\int_K \left(\left(\frac{\partial \varphi_3}{\partial x^2} - \frac{\partial \varphi_2}{\partial x^3} \right) dx^2 \wedge dx^3 + \left(\frac{\partial \varphi_1}{\partial x^3} - \frac{\partial \varphi_3}{\partial x^1} \right) dx^3 \wedge dx^1 \right.$$

$$\left. + \left(\frac{\partial \varphi_2}{\partial x^1} - \frac{\partial \varphi_1}{\partial x^2} \right) dx^1 \wedge dx^2 \right) = \int_{\partial K} \sum_{i=1}^{3} \varphi_i \, dx^i. \tag{2.125}$$

Equations (2.113) and (2.109) give the interpretation of the integrals in (2.125). So we see that (2.125) is nothing but the good old Stokes theorem.

(d) $p = 1$. Let $x^1, x^2, \ldots, x^n$ be Cartesian coordinates on $\mathcal{U} = \mathbb{R}^n$. The 'surface' K is now a curve with parameter representation

$$Q : [a, b] \to \mathcal{U}, \quad a < b,$$
$$\tau \mapsto Q(\tau). \tag{2.126}$$

The orientation of K is given by the 1-form $d\tau$ on the parameter space. For the 0-form, i.e. function, $\varphi \in \Lambda^0 \mathcal{U}$ we have

$$d\varphi = \sum_{i=1}^{n} \frac{\partial \varphi}{\partial x^i} \, dx^i. \tag{2.127}$$

With the proper interpretation of the orientation of a single point we get from Stokes' theorem

$$\int_K \sum_{i=1}^n \frac{\partial \varphi}{\partial x^i} dx^i = \varphi(Q(b)) - \varphi(Q(a)), \qquad (2.128)$$

again a well-known formula if we remember that the lhs of this equation is the line integral of grad φ along K. In the special case $n = 1$ with $Q = $ inclusion map, we recover from (2.128)

$$\int_a^b \frac{d\varphi}{dx} dx = \varphi(b) - \varphi(a), \qquad (2.129)$$

i.e. the fundamental theorem of calculus.

Note that the relations (2.121), (2.123), (2.125), and (2.128) are valid for arbitrary coordinates x^i, since Stokes' theorem (2.118) is coordinate-free. The use of Cartesian coordinates, which we assumed in all the above examples, is, however, necessary for the interpretation in terms of the operations of vector analysis.

Problems

2.1 Derive the formulas (2.14) and (2.29) for the transformation of frames associated with coordinate systems.

2.2 Prove that pullback and exterior derivative commute (equation (2.63)).

2.3 Verify that the function (2.76) satisfies the equation $df = \varphi$.

2.4 Prove Stokes' theorem (2.118) with $p = n = 2$ in the special case where K is a rectangle.

3

Metric structures

In this chapter we introduce a metric on our open subset of $\mathbb{R}^n$. For the moment let us consider a surface that can be parametrized by two coordinates and thereby be viewed as a subset of $\mathbb{R}^2$. We should not imagine it flat – nor curved. An adequate model is a rubber sheet where lengths and angles are undefined. What the surface inherits from $\mathbb{R}^2$ is its topology, those properties which do not change when the sheet is smoothly stretched or otherwise deformed. Typical properties are: dimensionality, connectedness, holes. Note that the dimensionality of a hole has no meaning. For example if we cut a point, 'zero-dimensional hole', out of our sheet and then stretch it, the hole can become one- or two-dimensional.

Technically, a deformation is given by a diffeomorphism F (active transformation). A vector field can be thought of as little arrows drawn on the rubber sheet. Deformations naturally act on vector fields by displacing the base points of the arrows and by turning and stretching or shrinking them. This action was named tangent mapping $T_x F$ in chapter 2.

A metric fixes lengths and angles on that rubber sheet. It becomes rigid, but not in the sense of being plastered, because this would imply an embedding of the sheet in our three-dimensional space, thereby also fixing angles and lengths in directions normal to the surface.

An adequate model of a surface with metric is a sheet of paper. It cannot be stretched in its plane. Deformations perpendicular to the sheet like rolling it up are irrelevant: A sheet of paper is an example of a flat space – the three angles of a triangle always add up to 180° – whether the sheet is rolled or not.

Now let us imagine a curved surface, for instance Europe considered as a piece of a two-dimensional sphere. How do we measure a length? We could go to Paris and have a look at the standard meter. Being a straight bar it is not really in the surface. Indeed we shall define a metric in the tangent space, here T_{Paris} Europe, which is a vector space. This definition has a priori nothing to do with the axioms of a metric space where a distance is defined directly in the space itself. We begin by reviewing metric structures on vector spaces.

3.1 Pseudometric on vector spaces

Let V be a real n-dimensional vector space. A pseudometric – the prefix pseudo will often be dropped – on V is by definition a bilinear form

$$g: V \times V \to \mathbb{R}$$

$$(v, w) \mapsto g(v, w)$$

which is symmetric,

$$g(v, w) = g(w, v), \tag{3.1}$$

and nondegenerate; i.e., $g(v, w) = 0$ for all $w \in V$ implies $v = 0$.

The metric being bilinear, it is sufficient to know its values on a basis. Let b_i, $i = 1, 2, \ldots, n$, be a basis of V. We define the matrix of the metric with respect to this basis as

$$g_{ij} := g(b_i, b_j). \tag{3.2}$$

It is a symmetric $n \times n$ matrix with nonvanishing determinant:

$$\det(g_{ij}) \neq 0. \tag{3.3}$$

Under a change of basis described by a matrix $\gamma \in GL_n$,

$$b_i' = \sum_j (\gamma^{-1})^j{}_i b_j, \tag{3.4}$$

the matrix of the metric transforms as

$$g_{ij}' := g(b_i', b_j') = (\gamma^{-1T} g \gamma^{-1})_{ij}. \tag{3.5}$$

THEOREM (Gram–Schmidt)

Any metric admits an orthonormal basis, i.e. a basis $e_1, \ldots, e_n$ such that

$$g(e_i, e_j) = \eta_{ij}, \tag{3.6}$$

where η is a diagonal matrix with r plus ones and s minus ones, $r + s = n$:

$$\eta := \mathrm{diag}\left(\underbrace{1, 1, \ldots, 1}_{r}, \underbrace{-1, \ldots, -1}_{s} \right). \tag{3.7}$$

In other words, for any symmetric $n \times n$ matrix g with nonvanishing determinant there is an invertible matrix γ such that

$$\gamma^{-1T} g \gamma^{-1} = \eta. \tag{3.8}$$

Note that an orthonormal basis is not unique. We define the pseudoorthogonal group $O(r,s)$ as the subgroup of GL_n consisting of matrices Λ leaving η invariant:

$$\Lambda^{-1T}\eta\Lambda^{-1} = \eta. \tag{3.9}$$

Now any change of basis described by a matrix Λ gives, starting from an orthonormal basis, again an orthonormal basis, and conversely any two orthonormal bases are connected by such a transformation Λ.

THEOREM (Sylvester)

The integer r does not depend on the choice of the orthonormal basis.

The difference between the number of plus ones and the number of minus ones, $r - s$, is often called the signature of the metric. If $s = 0$, the metric is positive-definite: $g(v,v) \geqslant 0$ for all $v \in V$, $g(v,v) = 0$ implies $v = 0$. Strictly speaking only then is it called a metric or scalar product, otherwise it is a pseudometric. We shall not insist on this distinction.

A metric can also be defined by choosing a basis e_i and declaring it to be orthonormal:

$$g(e_i, e_j) := \eta_{ij}. \tag{3.10}$$

Of course, then two sets of basis vectors connected by a pseudoorthogonal transformation Λ define the same metric.

3.2 Induced metric on the dual space

Let e_i, $i = 1, 2, \ldots, n$, be an orthonormal basis of V and e^i the corresponding dual basis of V^*. We define a metric $\overset{*}{g}$ on V^* by declaring this basis to be orthonormal:

$$\overset{*}{g}(e^i, e^j) := \eta_{ij}. \tag{3.11}$$

One easily verifies that this metric does not depend on the choice of the orthonormal basis e_i of V.

If V is oriented and the basis e_i is an oriented basis, the n-form $e^1 \wedge e^2 \wedge \cdots \wedge e^n$ is the volume form: Evaluation of this form on n vectors gives the signed volume of the parallelepiped spanned by the vectors.

For an arbitrary basis b_i of V and corresponding dual basis β^i of V^* we define the matrix of the metric $\overset{*}{g}$ by

$$g^{ij} := \overset{*}{g}(\beta^i, \beta^j). \tag{3.12}$$

Note that this matrix is distinguished from its analogue in V only by the position of the indices. A short calculation shows that one is the inverse of the other:

$$\sum_k g^{ik} g_{kj} = \delta^i{}_j. \qquad (3.13)$$

In particular, η is its own inverse.

V and V^* are both n-dimensional vector spaces. Therefore they are isomorphic:

$$V \overset{\cong}{\rightarrow} V^*$$
$$b_i \mapsto \beta^i.$$

However, this isomorphism is noncanonical; i.e. it depends on the choice of the basis b_i.

The metric being nondegenerate induces a canonical isomorphism

$$J: V \overset{\cong}{\rightarrow} V^*$$
$$v \mapsto g(v, \cdot),$$

which in physics is called 'lowering and raising indices with g_{ij} and g^{ij}'. For expressed in a basis this isomorphism reads

$$J(b_i) = \sum_j g_{ij} \beta^j. \qquad (3.14)$$

3.3 The Hodge star

Our starting point is the observation that

$$\dim \Lambda^p V = \dim \Lambda^{n-p} V = \binom{n}{p}. \qquad (3.15)$$

With the help of an orientation ω and a metric g we can define a canonical isomorphism $*$: Let e^i be an oriented orthonormal basis of $V^* = \Lambda^1 V$. The Hodge star is the linear map

$$*: \Lambda^p V \rightarrow \Lambda^{n-p} V$$

given by

$$*(e^{i_1} \wedge \cdots \wedge e^{i_p}) := \varepsilon_{i_1 \cdots i_n} \eta^{i_1 i_1} \cdots \eta^{i_p i_p} e^{i_{p+1}} \wedge \cdots \wedge e^{i_n}. \qquad (3.16)$$

We have to convince ourselves that this definition does not depend on the choice of the oriented orthonormal basis. We remark that any two oriented orthonormal bases are connected by a transformation Λ which is pseudoorthogonal and has unit determinant,

$$\det \Lambda = 1. \tag{3.17}$$

The set of all such matrices forms a subgroup of $O(r, s)$ called special pseudoorthogonal group $SO(r, s)$. Now equation (3.16) behaves well under this group, the S, equation (3.17), takes care of $\varepsilon_{i_1 \cdots i_n}$ and the O, equation (3.9), of the ηs.

The linear map $*$ is bijective; indeed its square is plus or minus the identity:

$$**\varphi = (-1)^{p(n-1)+s}\varphi, \quad \varphi \in \Lambda^p V. \tag{3.18}$$

In an arbitrary oriented basis β^i any p-form φ can be written

$$\varphi =: \frac{1}{p!} \sum_{i_1,\ldots,i_p} \varphi_{i_1 \cdots i_p} \beta^{i_1} \wedge \cdots \wedge \beta^{i_p}. \tag{3.19}$$

It follows that

$$*\varphi = \frac{1}{(n-p)!} \sum_{i_{p+1},\ldots,i_n} \left[\frac{1}{p!} \sum_{i_1,\ldots,i_p} \varepsilon_{i_1 \cdots i_n} |\det g_{kl}|^{1/2} \right.$$

$$\left. \times \sum_{j_1,\ldots,j_p} \varphi_{j_1 \ldots j_p} g^{i_1 j_1} \ldots g^{i_p j_p} \right] \beta^{i_{p+1}} \wedge \cdots \wedge \beta^{i_n}. \tag{3.20}$$

Therefore, in tensor language, $*$ amounts to raising all indices of $\varphi_{i_1 \cdots i_p}$ and then contracting it with the ε-tensor.

We briefly mention two examples from physics. In $\mathbb{R}^3$ with Euclidean metric ($\eta = 1$) an infinitesimal rotation is an antisymmetric matrix and can be represented by a 2-form. Its star is a 1-form. The canonical isomorphism J makes it a vector, which is often used to describe the rotation. Similarly the magnetic field vector is arranged into an antisymmetric 3×3 matrix (F_{ij}). Note that both vectors are axial vectors transforming with a minus sign under parity. For in the definition of the star we had to specify an orientation.

3.4 Isometries

We conclude this subchapter on linear algebra with the definition of isometries. Let V and V' be two vector spaces with metrics g and g', respectively. An isometry is by definition a linear mapping

$$F: V \to V'$$

preserving the metric:

$$g'(Fv, Fw) = g(v, w) \tag{3.21}$$

for all v and w in V.

For example, if both vector spaces are $\mathbb{R}^4$ with Minkowski metric, $\eta = \text{diag}(1, -1, -1, -1)$, the isometries are just the Lorentz group $O(1, 3)$. Another example is the canonical isomorphism

$$J: V \to V^*.$$

It is an isometry with respect to g and $\overset{*}{g}$.

3.5 Metric structures on an open subset of $\mathbb{R}^n$

To an open subset $\mathscr{U}$ of $\mathbb{R}^n$ we have associated a collection of vector spaces $T_x \mathscr{U}$ indexed by the points in $\mathscr{U}$. We define a metric on $\mathscr{U}$ by a collection of vector space metrics,

$$g_x: T_x \mathscr{U} \times T_x \mathscr{U} \to \mathbb{R},$$

varying smoothly with x.

To be more explicit, we choose a frame b_i, $i = 1, \ldots, n$. Now we have in each point x a basis of $T_x \mathscr{U}$ and as before we set

$$g_{ij}(x) := g_x(b_i(x), b_j(x)). \tag{3.22}$$

In other words: With respect to a given frame a metric on $\mathscr{U}$ is represented by a symmetric $n \times n$ matrix, the coefficients of which are smooth functions of x, with nowhere vanishing determinant.

Under a change of frame

$$b_i'(x) = \sum_j (\gamma^{-1}(x))^j{}_i b_j(x) \tag{3.23}$$

we have pointwise the same formulas as before:

$$g_{ij}'(x) = (\gamma^{-1T}(x) g(x) \gamma^{-1}(x))_{ij}. \tag{3.24}$$

For example, we consider on $\mathscr{U} = \mathbb{R}^3 - x^3$-axis the frames associated with Cartesian and polar coordinates:

$$b_1 = \frac{\partial}{\partial x^1}, \quad b_2 = \frac{\partial}{\partial x^2}, \quad b_3 = \frac{\partial}{\partial x^3}, \tag{3.25}$$

$$b_1' = \frac{\partial}{\partial r}, \quad b_2' = \frac{\partial}{\partial \varphi}, \quad b_3' = \frac{\partial}{\partial \theta}, \tag{3.26}$$

$$x^1 = r \cos \varphi \sin \theta,$$
$$x^2 = r \sin \varphi \sin \theta, \tag{3.27}$$
$$x^3 = r \cos \theta,$$

for example

$$\frac{\partial}{\partial r} = \frac{\partial x^1}{\partial r} \frac{\partial}{\partial x^1} + \frac{\partial x^2}{\partial r} \frac{\partial}{\partial x^2} + \frac{\partial x^3}{\partial r} \frac{\partial}{\partial x^3} \tag{3.28}$$

and

$$\gamma^{-1} = \begin{pmatrix} \cos \varphi \sin \theta & -r \sin \varphi \sin \theta & r \cos \varphi \cos \theta \\ \sin \varphi \sin \theta & r \cos \varphi \sin \theta & r \sin \varphi \cos \theta \\ \cos \theta & 0 & -r \sin \theta \end{pmatrix}. \tag{3.29}$$

The Euclidean metric on $\mathbb{R}^3$ is defined by the identity matrix 1 with respect to the Cartesian frame b_i:

$$(g_{ij}) = \begin{pmatrix} 1 & 0 & 0 \\ 0 & 1 & 0 \\ 0 & 0 & 1 \end{pmatrix}. \tag{3.30}$$

With respect to the polar frame b_i' we obtain

$$(g_{ij}') = \gamma^{-1T} 1 \gamma^{-1} = \begin{pmatrix} 1 & 0 & 0 \\ 0 & r^2 \sin^2 \theta & 0 \\ 0 & 0 & r^2 \end{pmatrix}. \tag{3.31}$$

3.6 Holonomic and orthonormal frames

In many situations it is convenient to work with dual frames. In our example

$$\beta^1 = dx^1, \quad \beta^2 = dx^2, \quad \beta^3 = dx^3, \tag{3.32}$$

$$\beta'^1 = dr, \quad \beta'^2 = d\varphi, \quad \beta'^3 = d\theta, \tag{3.33}$$

$$(g^{ij}) = \begin{pmatrix} 1 & 0 & 0 \\ 0 & 1 & 0 \\ 0 & 0 & 1 \end{pmatrix}, \tag{3.34}$$

$$(g'^{ij}) = \begin{pmatrix} 1 & 0 & 0 \\ 0 & \dfrac{1}{r^2 \sin^2 \theta} & 0 \\ 0 & 0 & \dfrac{1}{r^2} \end{pmatrix} \tag{3.35}$$

A dual frame β^i can have two desirable properties:

(a) It is called holonomic if the n 1-forms β^i are exact, i.e. if there are 0-forms (functions) x^i such that

$$\beta^i = dx^i. \tag{3.36}$$

The linear independence of the β^i then ensures locally that the x^i are coordinates and $\beta^1, \beta^2, \ldots, \beta^n$ is the frame associated with these coordinates. Of course there are always (global) coordinates on an open subset of $\mathbb{R}^n$ and therefore also holonomic frames.

(b) A frame e^i is orthonormal if

$$\mathring{g}_x(e^i(x), e^j(x)) = \eta_{ij}. \tag{3.37}$$

It is always possible to find an orthonormal frame. Just apply in all points the orthonormalization procedure by Gram and Schmidt to an arbitrary frame. The resulting orthonormal frame will be differentiable because this procedure involves only addition, multiplication, and division.

Naturally one would like to have both properties together. We shall see later that there exists a holonomic and orthonormal frame if and only if $\mathscr{U}$ is flat, e.g. dx, dy, dz for Euclidean $\mathbb{R}^3$.

As a nonflat example let us take a piece of the sphere $r = 1$ in $\mathbb{R}^3$. We consider it as a subset of $\mathbb{R}^2$ parametrized by φ and θ, say $20° < \varphi < 30°$, $30° < \theta < 40°$; $d\varphi$, $d\theta$ is a holonomic frame. We define the metric with respect to this frame (see (3.35)):

$$\begin{pmatrix} \dfrac{1}{\sin^2 \theta} & 0 \\ 0 & 1 \end{pmatrix}.$$

An orthonormal frame is $\sin \theta \, d\varphi$, $d\theta$. It is not holonomic because

$$d(\sin \theta \, d\varphi) = \cos \theta \, d\theta \wedge d\varphi \neq 0. \tag{3.38}$$

In general it is not easy to decide whether a given metric admits a holonomic orthonormal frame. We shall see later an algorithm to answer this question when we introduce the curvature.

As already mentioned, comoving frame, tetrad, vielbein are synonyms for frame. Some authors reserve one word or the other for orthonormal frames.

3.7 Isometries of open subsets of $\mathbb{R}^n$

We define isometries for an open subset $\mathcal{U}$ of $\mathbb{R}^n$ endowed with a metric g. Let $F: \mathcal{U} \to \mathcal{U}$ be a diffeomorphism. It is called an isometry if for all x in $\mathcal{U}$ the tangent mapping

$$T_x F: T_x \mathcal{U} \to T_{F(x)} \mathcal{U}$$

is an isometry with respect to the vector space metrics g_x and $g_{F(x)}$.

Note that the isometries of an open subset form a group. If $\mathcal{U}$ is the entire $\mathbb{R}^4$ with Minkowski metric, then this group is the Poincaré group.

3.8 Coderivative

Let $\mathcal{U}$ be an open subset of $\mathbb{R}^n$ with an orientation ω and a metric g of signature $r - s$. We combine the algebraic operation $*$ with the exterior derivative to define the coderivative δ:

$$\delta: \Lambda^p \mathcal{U} \to \Lambda^{p-1} \mathcal{U}$$

$$\varphi \mapsto \delta \varphi,$$

$$\delta \varphi := (-1)^{np+n+1+s} * \mathrm{d} * \varphi. \qquad (3.39)$$

Like the exterior derivative, the coderivative is a first-order differential operator, but contrary to d it lowers the degree of differential forms by one unit.

Applied to a 1-form $\varphi = \sum_{i=1}^n \varphi_i \, \mathrm{d}x^i$ it yields the function

$$\delta \varphi = \frac{-1}{|g|^{1/2}} \sum_{k,j=1}^n \frac{\partial}{\partial x^k} (|g|^{1/2} \varphi_j g^{kj}) \qquad (3.40)$$

with the standard abbreviation

$$g(x) := \det(g_{ij}(x)). \qquad (3.41)$$

Note that g is not simply a function on $\mathcal{U}$ (0-form), it also depends on the frame used to define the matrix g_{ij}.

Two properties of the coderivative follow immediately:

$$\delta^2 = 0, \qquad (3.42)$$

$$\int d\varphi \wedge *\psi = \int \varphi \wedge *\delta\psi, \quad \varphi \in \Lambda^{p-1}\mathcal{U}, \quad \psi \in \Lambda^p\mathcal{U}. \tag{3.43}$$

The last equation holds up to the boundary term $\int d(\varphi \wedge *\psi)$, which vanishes if we take, for instance, φ with compact support. Recall that the support of a function is the closure of the set of points where this function is different from zero.

Equation (3.43) means: The coderivative is the formal adjoint of the exterior derivative with respect to the 'scalar product' of p-forms $\int \varphi \wedge *\psi$. In components it reads:

$$\int \varphi \wedge *\psi = \frac{1}{p!} \sum_{\substack{i_1,\dots,i_p \\ j_1,\dots,j_p}} \int \varphi_{i_1\dots i_p} \psi_{j_1\dots j_p}$$

$$\times g^{i_1 j_1} \cdots g^{i_p j_p} |g|^{1/2} \, dx^1 \wedge \cdots \wedge dx^n. \tag{3.44}$$

In particular,

$$\int \varphi \wedge *\psi = \int \psi \wedge *\varphi. \tag{3.45}$$

Next we define the second-order differential operator

$$\Delta: \Lambda^p\mathcal{U} \to \Lambda^p\mathcal{U}$$
$$\Delta := -(d\delta + \delta d). \tag{3.46}$$

If the metric is positive-definite, $s = 0$, it is called Laplacian. If $r = 1$, it is called the d'Alembert operator and is usually written $\square$. On 0-forms f, Δ has the well-known form

$$\Delta f = \frac{1}{|g|^{1/2}} \sum_{k,j=1}^{n} \frac{\partial}{\partial x^k} \left[|g|^{1/2} \left(\frac{\partial}{\partial x^j} f \right) g^{kj} \right]. \tag{3.47}$$

Finally we write out gradient, divergence, and curl in coordinate-free manner: Let f be a function. Its gradient is the vector field

$$\operatorname{grad} f := J^{-1} df = \sum_{i,j=1}^{n} \left(\frac{\partial}{\partial x^j} f \right) g^{ji} \frac{\partial}{\partial x^i}, \tag{3.48}$$

where J maps vector fields onto 1-forms by pointwise application of the canonical isometry (3.14). The divergence of a vector field $v = \sum_{i=1}^{n} v^i (\partial/\partial x^i)$ is the function

$$\operatorname{div} v := -\delta J v = \frac{1}{|g|^{1/2}} \sum_{k=1}^{n} \frac{\partial}{\partial x^k} (|g|^{1/2} v^k). \tag{3.49}$$

The curl is defined only in $\mathbb{R}^3$:

$$\operatorname{curl} v := J^{-1} * dJv = \frac{1}{|g|^{1/2}} \sum_{i,l,j,k=1}^{3} \left[\frac{\partial}{\partial x^j} (v^l g_{li}) \varepsilon_{jik} \right] \frac{\partial}{\partial x^k}. \tag{3.50}$$

3.9 Summarizing chapters 1, 2, 3

We summarize chapters 1, 2, 3 in the following display indicating the correspondence between index-free and component notations:

$\varphi \in \Lambda^p \mathscr{U}$ $\varphi_{[i_1 \cdots i_p]}$

$\varphi \wedge \psi$ $\varphi_{[i_1 \cdots i_p} \psi_{j_1 \cdots j_q]}$

$v \in T_x \mathscr{U}$ v^i

$i_v \varphi$ $\sum_j v^j \varphi_{j i_1 \cdots i_{p-1}}$

$d\varphi$ $\partial_{[i_1} \varphi_{i_2 \cdots i_{p+1}]}$

$\displaystyle\int_K \varphi$ $\displaystyle\int_K \varphi_{1 \cdots n} \, dx^1 \cdots dx^n$

g $g_{ij}, \quad \det(g_{ij}) \neq 0$

$\overset{*}{g}$ $g^{ij} = (g^{-1})_{ij}$

$\varphi = J(v)$ $\varphi_i = \sum_j g_{ij} v^j$

$*\varphi$ $\displaystyle\sum_{\substack{j_1, \ldots, j_p \\ i_1, \ldots, i_p}} |g|^{1/2} \varphi_{j_1 \cdots j_p} g^{i_1 j_1} \cdots g^{i_p j_p} \varepsilon_{i_1 \cdots i_p i_{p+1} \cdots i_n}$

$-d*d* - *d*d$ $\square$

Problems

3.1 Verify the transformation law (3.5) of the matrix describing a metric under a change of basis.

3.2 Calculate the square of the Hodge star.

3.3 Derive the expression (3.20) of the Hodge star with respect to an arbitrary basis.

3.4 Verify the expression (3.47) of the Laplace operator applied to a 0-form.

4
Gauge theories

Let us now see how we can apply the mathematics treated so far in the formulation of gauge theories. We shall discuss the abelian case (Maxwell's equations), the geometry of (local) gauge invariance, and nonabelian (Yang–Mills) theories restricting ourselves to four-dimensional Minkowski space. The generalization to dimensions different from four is straightforward. Furthermore, all formulas written in terms of differential forms will remain valid without modification in any curved space with Minkowski signature.

4.1 Maxwell's equations

In standard notation, Maxwell's equations read:

$$\operatorname{curl} E + \frac{\partial}{\partial t} B = 0, \quad \operatorname{div} B = 0,$$

$$\operatorname{curl} H - \frac{\partial}{\partial t} D = j, \quad \operatorname{div} D = \rho. \tag{4.1}$$

They are completed by the relations

$$B = H, \quad D = E. \tag{4.2}$$

Here E is the electric field, B the magnetic induction, H the magnetic field, and D the electric displacement. The charge density is denoted by ρ and the density of the electric current by j. We use Heaviside units and put the velocity of light in the vacuum equal to one.

We want to describe electromagnetism by differential forms on spacetime. Our spacetime is Minkowski space: $\mathbb{R}^4$ with Cartesian coordinates x^0, x^1, x^2, x^3 as an oriented coordinate system and metric

$$g\left(\frac{\partial}{\partial x^\mu}, \frac{\partial}{\partial x^\nu}\right) = \eta_{\mu\nu}, \quad \eta = \operatorname{diag}(1, -1, -1, -1). \tag{4.3}$$

The coordinate x^0 is the time t, and we follow the popular convention in labelling spacetime coordinates by Greek indices. These are raised and lowered with the help of the metric tensor η. Sometimes we shall employ Einstein's summation convention: An index which appears twice in an expression, once above and once below, is to be summed over.

We combine all electromagnetic fields into two 2-forms F and G:

$$F := -\sum_{i=1}^{3} E_i \, dx^0 \wedge dx^i + \tfrac{1}{2} \sum_{i,j,k=1}^{3} \varepsilon_{ijk} B_i \, dx^j \wedge dx^k$$

$$= \tfrac{1}{2} F_{\mu\nu} \, dx^\mu \wedge dx^\nu, \tag{4.4}$$

$$G := \sum_{i=1}^{3} H_i \, dx^0 \wedge dx^i + \tfrac{1}{2} \sum_{i,j,k=1}^{3} \varepsilon_{ijk} D_i \, dx^j \wedge dx^k$$

$$= \tfrac{1}{2} G_{\mu\nu} \, dx^\mu \wedge dx^\nu. \tag{4.5}$$

Explicitly, the matrices of the coefficients $F_{\mu\nu}$ and $G_{\mu\nu}$ read:

$$(F_{\mu\nu}) = \begin{pmatrix} 0 & -E_1 & -E_2 & -E_3 \\ E_1 & 0 & B_3 & -B_2 \\ E_2 & -B_3 & 0 & B_1 \\ E_3 & B_2 & -B_1 & 0 \end{pmatrix}, \tag{4.6}$$

$$(G_{\mu\nu}) = \begin{pmatrix} 0 & H_1 & H_2 & H_3 \\ -H_1 & 0 & D_3 & -D_2 \\ -H_2 & -D_3 & 0 & D_1 \\ -H_3 & D_2 & -D_1 & 0 \end{pmatrix}. \tag{4.7}$$

The charge density and the components of the current density are collected into the current density 4-vector j^μ, where $j^0 = \rho$. This 4-vector, in turn, is used to construct the current density 3-form j:

$$j = \frac{1}{3!} \varepsilon_{\mu\nu\lambda\rho} j^\mu \, dx^\nu \wedge dx^\lambda \wedge dx^\rho. \tag{4.8}$$

It is natural to associate a 3-form with the current density and not a 1-form. For current and charge densities are objects which after integration over a three-dimensional volume in spacetime give the charge contained in that volume.

It is now easy to check that Maxwell's equations can be written as

$$dF = 0, \tag{4.9}$$

$$dG = j. \tag{4.10}$$

In this form, they are coordinate-free and do not even involve a metric.

Consequently, equations (4.9), (4.10) with $j = 0$ are invariant under arbitrary coordinate transformations, not just Poincaré transformations. What restricts the invariance group of electromagnetism to the Poincaré group are the equations (4.2), which in terms of our forms read

$$*F = G. \tag{4.11}$$

For the definition of the Hodge star $*$ we need a metric. Therefore, (4.11) is invariant only under the isometries of Minkowski space, i.e. the Poincaré group.

The current density j in (4.10) cannot be chosen arbitrarily. Because $d^2 = 0$, (4.10) implies

$$dj = 0; \tag{4.12}$$

i.e. Maxwell's equations are consistent only if (4.12) is fulfilled. Since

$$dj = \partial_\mu j^\mu \, dx^0 \wedge dx^1 \wedge dx^2 \wedge dx^3, \tag{4.13}$$

we see that (4.12) expresses current conservation.

Eliminating G with the help of (4.11), we find

$$dF = 0, \tag{4.14}$$

$$\delta F = *j, \tag{4.15}$$

because in our case $\delta = *d*$. So F is closed, and in star-shaped regions the Poincaré lemma guarantees the existence of a potential 1-form A such that F can be written as

$$F = dA. \tag{4.16}$$

Of course, F determines A only up to a gauge transformation: For any 0-form Λ, A and $A + d\Lambda$ lead to the same field strength F. Whereas the homogeneous Maxwell equations, $dF = 0$, are identically satisfied by (4.16), the inhomogeneous equations, $\delta F = *j$, give us an equation of motion for A:

$$\delta dA = *j. \tag{4.17}$$

If we choose, for example, the Lorentz gauge,

$$\delta A = 0, \tag{4.18}$$

we can write

$$(\delta d + d\delta)A = *j \tag{4.19}$$

or

$$\Box A = - *j. \tag{4.20}$$

The equation of motion (4.17) may be derived from the action

$$S[A] = \int(-\tfrac{1}{2}F \wedge *F - j \wedge A)$$

$$= \int(-\tfrac{1}{4}F_{\mu\nu}F^{\mu\nu} + j_\mu A^\mu)\,dx^0 \wedge dx^1 \wedge dx^2 \wedge dx^3. \quad (4.21)$$

Note that this action is gauge invariant only for a conserved current j:

$$S[A + d\Lambda] - S[A] = -\int j \wedge d\Lambda = \int d(j \wedge d\Lambda) - \int (dj) \wedge \Lambda. \quad (4.22)$$

The first term vanishes due to Stokes' theorem (up to boundary contributions), whereas the second term is equal to zero for arbitrary Λ only if $dj = 0$.

Let us derive (4.17) from the condition of stationarity of the action (4.21). Doing this we have to consider S as a functional of A; i.e. F in (4.21) is merely an abbreviation of dA. We write the variation of A as

$$A \to A + a \quad (4.23)$$

and find keeping only those terms linear in a:

$$S[A + a] - S[A] = \int(-(da) \wedge *dA + a \wedge j) + \cdots \quad (4.24)$$

Here we have used (3.45). Applying (3.43) and ignoring boundary terms we arrive at

$$S[A + a] - S[A] = \int a \wedge *(-\delta dA + *j) + \cdots \quad (4.25)$$

So S is stationary if (4.17) is fulfilled.

In order to make the analogy of the $U(1)$ gauge theory of electromagnetism with nonabelian gauge theories manifest, we introduce a coupling constant g and define

$$\tilde{A} := -igA, \quad \tilde{F} := -igF, \quad \tilde{j} := -(i/g)j. \quad (4.26)$$

In terms of these (admittedly odd) variables, the Lagrangian of electrodynamics reads

$$\mathcal{L} = \frac{1}{2g^2}\tilde{F} \wedge *\tilde{F} + \tilde{j} \wedge \tilde{A}. \quad (4.27)$$

A gauge transformation of the old potential,

$$A \to A + d\Lambda, \tag{4.28}$$

may be written as

$$\tilde{A} \to e^{\tilde{\Lambda}} \tilde{A} e^{-\tilde{\Lambda}} + e^{\tilde{\Lambda}} d e^{-\tilde{\Lambda}}, \tag{4.29}$$

where $\tilde{\Lambda} = ig\Lambda$. Note that $e^{\tilde{\Lambda}}$ is a function on spacetime with values in the group $U(1)$. In the following, the notation will be such that A and F correspond to $\tilde{A}$ and $\tilde{F}$ if one specializes to the $U(1)$ case.

4.2 Connection = potential

The Lagrangian (4.27) of electrodynamics is invariant under the local $U(1)$-transformations (4.29). Invariance under local G-transformations, where G can be an arbitrary Lie group, characterizes gauge theories in general, to which we now turn. We assume that the reader has some basic knowledge about Lie groups, Lie algebras, group representations, etc. For the purpose of this chapter it is, however, sufficient to think of matrix groups like, for example, $SU(N)$. Precise definitions will be given in chapter 8.

So let G be a Lie group, $\mathfrak{g}$ its Lie algebra, and V an r-dimensional (real or complex) vector space in which a representation of G and $\mathfrak{g}$ acts. For the sake of simplicity we restrict ourselves to scalar matter fields, although this restriction is not essential. (Fermion fields will be dealt with in chapter 11.) We interpret a scalar field Φ as a 0-form with values in V:

$$\Phi: \mathcal{U} \to V. \tag{4.30}$$

As usual, $\mathcal{U}$ is an open subset of $\mathbb{R}^4$. Under a local gauge transformation, given by the function

$$\gamma: \mathcal{U} \to G, \tag{4.31}$$

Φ is transformed into Φ':

$$\Phi'(x) = \rho(\gamma(x))\Phi(x), \tag{4.32}$$

where

$$\rho: G \to GL(V) \tag{4.33}$$

is the representation of G on V. In the following, we shall write instead of (4.32)

$$\Phi'(x) = \gamma(x)\Phi(x) \tag{4.34}$$

and implicitly assume that the appropriate representation is taken. With respect to some basis of V the elements of G are identified with matrices. If G is compact, we shall use only bases such that these matrices are unitary. Consequently, $\mathfrak{g}$ is considered as consisting of antihermitian matrices for compact G.

Our goal is to investigate the geometry of gauge invariance, i.e. invariance under transformations of the type (4.34). Since γ is an arbitrary (smooth) function, $\Phi(x)$ transforms independently of $\Phi(y)$ ($y \neq x$). This fact may be interpreted in the following way. We have a family of vector spaces V_x (one for each $x \in \mathcal{U}$), which are isomorphic but not identical to V, such that the values of Φ at x lie in V_x. Nevertheless, we can consider $\Phi(x)$ as an element of V, as we did above (and shall often do in the sequel), after choosing bases of V and of each V_x: One identifies a vector in V_x with a vector in V, if they have the same expansion coefficients with respect to these bases. Changing the bases of the V_x will change $\Phi(x) \in V$ according to the gauge transformation (4.34). If G is not the whole linear group of V, this means that we admit only a restricted set of transformations of the bases.

All this is exactly as in the case of the tangent spaces of an open subset $\mathcal{U} \subset \mathbb{R}^n$. There we have at each $x \in \mathcal{U}$ the tangent space $T_x\mathcal{U}$ isomorphic to $\mathbb{R}^n$. A frame $b_1, b_2, \ldots, b_n$ on $\mathcal{U}$ provides us with an isomorphism of $T_x\mathcal{U}$ onto $\mathbb{R}^n$ mapping $v \in T_x\mathcal{U}$ onto its expansion coefficients with respect to the basis $b_1(x), b_2(x), \ldots, b_n(x)$ of $T_x\mathcal{U}$, and we can represent a vector field $v \in \text{vect}(\mathcal{U})$ as an $\mathbb{R}^n$-valued function on $\mathcal{U}$ (cf. (2.11)):

$$x \mapsto (v^1(x), \ldots, v^n(x)).$$

A change of the frame as in (2.13) transforms this function analogously to (4.34). If we have a metric and want to conserve orthonormality of frames, we must restrict ourselves to orthogonal matrices $\gamma(x)$ in (2.13).

The collection of tangent spaces as well as the collection of the V_x is appropriately described in the language of fibre bundles (see chapter 9). In physical terms we may interpret the requirement of local gauge invariance (independence of the fields at different spacetime points) as expressing the absence of (instantaneous) action at a distance.

The independence of the different vector spaces V_x entails the impossibility of identifying elements of V_x with elements of V_y in a gauge covariant manner, i.e., independently of any choice of bases. Such an identification is, however, necessary if we want to define derivatives of Φ, e.g. in order to write down field equations. The naive expression for a derivative of Φ

in direction u,

$$\lim_{\varepsilon \to 0} \frac{1}{\varepsilon} (\Phi(x + \varepsilon u) - \Phi(x)), \tag{4.35}$$

makes no (gauge covariant) sense, because $\Phi(x + \varepsilon u)$ and $\Phi(x)$ originally lie in the different vector spaces $V_{x+\varepsilon u}$ and V_x, respectively.

Establishing a connection between the V_x is therefore a necessary additional input, which will finally lead to the introduction of gauge fields beside the matter fields. We define such a connection by the concept of parallel translation along a curve using the fact that G acts in V. Let C be a curve with starting point x and end point y. The parallel transporter along C is an element of G denoted by $\Gamma[C]$. By means of the representation ρ we can apply it to $v \in V$ considered as representing an element of V_x. So we get a vector $\Gamma[C]v \in V$, which is interpreted as resulting from parallel translation along C and is identified with an element of V_y.

Of course, these definitions will make sense only if they do not depend on the choice of the bases used for the identifications: Gauge transformations and parallel translation should commute. This is achieved by assigning a proper transformation law under gauge transformations to $\Gamma[C]$. In formulas: Take $v \in V$ sitting at x. Gauge transformation leads to $\gamma(x)v$. Subsequent parallel translation gives $\Gamma'[C]\gamma(x)v$, to be considered as an element of V_y. The prime indicates that the gauge transformed parallel transporter is used. Interchanging the order of the operations, we end up with $\gamma(y)\Gamma[C]v$. So our construction will be gauge covariant if

$$\Gamma'[C] = \gamma(y)\Gamma[C]\gamma^{-1}(x). \tag{4.36}$$

Since parallel translation should be continuous, we assume that $\Gamma[C]$ is close to identity for an infinitesimal curve C. So $\Gamma[C] - 1$ is approximately equal to an element of $\mathfrak{g}$. On the other hand, an infinitesimal curve is essentially determined by its tangent vector at the starting point x, i.e. an element of $T_x \mathcal{U}$. In this way, parallel translation leads to a linear mapping from $T_x \mathcal{U}$ into $\mathfrak{g}$, i.e. a $\mathfrak{g}$-valued 1-form A on $\mathcal{U}$, which is called a connection 1-form (also potential or gauge field). It is the generalization of the electromagnetic potential.

Let us be more precise and consider a curve C with parameter representation $Q(\tau)$ $(0 \leqslant \tau \leqslant 1)$. The tangent vector of C at $Q(\tau)$ is denoted by $\dot{Q}(\tau)$. Let $\Phi(\tau) \in V$ sitting at $Q(\tau)$ be the result of parallel translation of $\Phi(0)$ along the portion of C between $\tau = 0$ and τ. Then, of course,

$\Phi(\tau + \varepsilon)$ results from $\Phi(\tau)$ by parallel translation along the part of C between τ and $\tau + \varepsilon$, and for $\varepsilon \to 0$ we can read off the corresponding parallel transporter:

$$\Phi(\tau + \varepsilon) = \Phi(\tau) - \varepsilon A(\dot{Q}(\tau))\Phi(\tau) + O(\varepsilon^2)$$

$$= [\mathbb{1} - \varepsilon A(\dot{Q}(\tau))]\Phi(\tau) + O(\varepsilon^2). \tag{4.37}$$

Here $A(\dot{Q}(\tau))$ means the 1-form A at $Q(\tau)$ evaluated on the tangent vector $\dot{Q}(\tau)$ and would be written out in full as $A|_{Q(\tau)}(\dot{Q}(\tau))$. From (4.37) we get a differential equation describing parallel transport along C:

$$\frac{d}{d\tau}\Phi(\tau) = -A(\dot{Q}(\tau))\Phi(\tau). \tag{4.38}$$

Standard theorems on ordinary linear differential equations guarantee the existence of a unique solution to (4.38), once an initial value $\Phi(0)$ is chosen. This solution can be given explicitly by means of a path-ordered exponential (cf. (4.43) below).

How does A behave with respect to gauge transformations? Equation (4.36), with C replaced by the infinitesimal part of C between τ and $\tau + \varepsilon$, yields

$$\mathbb{1} - \varepsilon A'(\dot{Q}(\tau)) = \gamma(Q(\tau + \varepsilon))[\mathbb{1} - \varepsilon A(\dot{Q}(\tau))]\gamma^{-1}(Q(\tau)). \tag{4.39}$$

Expanding in powers of ε we find

$$\gamma(Q(\tau + \varepsilon)) = \gamma(Q(\tau)) + \varepsilon \frac{d}{d\tau}\gamma(Q(\tau)) + O(\varepsilon^2)$$

$$= \gamma(Q(\tau)) + \varepsilon \, d\gamma|_{Q(\tau)}(\dot{Q}(\tau)) + O(\varepsilon^2) \tag{4.40}$$

(cf. (2.47)) and consequently

$$A'(\dot{Q}(\tau)) = \gamma(Q(\tau))A(\dot{Q}(\tau))\gamma^{-1}(Q(\tau)) - (d\gamma(\dot{Q}(\tau)))\gamma^{-1}(Q(\tau))$$

$$= \gamma(Q(\tau))A(\dot{Q}(\tau))\gamma^{-1}(Q(\tau)) + \gamma(Q(\tau)) \, d\gamma^{-1}(\dot{Q}(\tau)). \tag{4.41}$$

Expressions like $(d\gamma(\dot{Q}(\tau)))\gamma^{-1}(Q(\tau))$ are to be understood in the following sense: Take the exterior derivative of the matrix valued function γ (component by component), evaluate it at $Q(\tau)$ on $\dot{Q}(\tau)$, and multiply the resulting matrix with $\gamma^{-1}(Q(\tau))$. By choosing different paths C one can arrange that the tangent vectors $\dot{Q}(\tau)$ in (4.41) cover the whole tangent space $T_{Q(\tau)}\mathcal{U}$. Therefore we conclude that the gauge transformation described by γ acts on the connection A according to

$$A \to A' = \gamma A \gamma^{-1} + \gamma \, d\gamma^{-1}. \tag{4.42}$$

Parallel translation along an infinitesimal piece of C is given by (4.37). Iterating this equation we get an expression for $\Gamma[C]$:

$$\Phi(1) = \lim_{N \to \infty} \left[1 - \frac{1}{N} A\left(\dot{Q}\left(\frac{N-1}{N} \right) \right) \right] \cdots$$

$$\left[1 - \frac{1}{N} A\left(\dot{Q}\left(\frac{1}{N} \right) \right) \right] \left[1 - \frac{1}{N} A(\dot{Q}(0)) \right] \Phi(0)$$

$$=: P \exp \left\{ - \int_C A \right\} \Phi(0) \qquad (4.43)$$

so that

$$\Gamma[C] = P \exp \left\{ - \int_C A \right\}. \qquad (4.44)$$

The path-ordered exponential $P \exp$ is defined by (4.43) and may be written as

$$P \exp \left\{ - \int_C A \right\} = \sum_{j=0}^{\infty} (-1)^j \int_0^1 d\tau_j \int_0^{\tau_j} d\tau_{j-1} \cdots$$

$$\int_0^{\tau_2} d\tau_1 A(\dot{Q}(\tau_j)) \cdots A(\dot{Q}(\tau_2)) A(\dot{Q}(\tau_1)), \qquad (4.45)$$

an expression which will be useful later on. (The path-ordering P is analogous to the time-ordering operation in quantum field theory.) If G is abelian, the order of the factors in (4.43) and (4.45) is, of course, irrelevant, and $P \exp$ reduces to the ordinary exponential function.

The transformation behaviour of the connection (4.42) was determined such that (4.36) holds for infinitesimal curves. From (4.39) it follows easily that $\Gamma[C]$ as given by (4.43) indeed satisfies (4.36) for arbitrary paths C:

$$P \exp \left\{ - \int_C A' \right\} = \gamma(Q(1)) P \exp \left\{ - \int_C A \right\} \gamma^{-1}(Q(0)). \qquad (4.46)$$

4.3 Curvature = field strength

For a given connection, $\Gamma[C]$ generally depends on the curve C: Parallel translation along two different paths connecting the same points x and y will usually not give the same result. Equivalently we can say that $\Gamma[C]$ may differ from the identity even if C is closed.

Let us examine the path dependence of $\Gamma[C]$ in greater detail. We take

C to be a closed curve and $S(C)$ a surface bounded by C. With the help of Stokes' theorem we can write in the case of electrodynamics ($G = U(1)$):

$$\Gamma[C] = \exp\left\{-\int_C A\right\} = \exp\left\{-\int_{S(C)} \mathrm{d}A\right\} = \exp\left\{-\int_{S(C)} F\right\}. \tag{4.47}$$

Equation (4.47) tells us: The deviation of $\Gamma[C]$ from the identity is determined by the integral of the field strength 2-form F over an arbitrary surface bounded by C. That this integral does not depend on the specific surface chosen follows from $\mathrm{d}F = 0$ via Stokes' theorem.

If G is nonabelian, the situation is complicated by the path-ordering in (4.44). Nevertheless, Schlesinger (1928) was able to derive an analogue of (4.47) for such groups (see also Dollard & Friedman 1979). We shall be content with a nonabelian generalization for infinitesimal curves C, because we only want to motivate the definition of a field strength for nonabelian gauge fields. So we assume that the closed curve C, given by a parameter representation $Q: [0, 1] \to \mathcal{U}$, is of infinitesimal length L. Let $S(C)$ be an infinitesimal surface (area of order L^2) bounded by C. We calculate $\Gamma[C]$ up to terms of order L^3.

Our starting point is equation (4.45):

$$P\exp\left\{-\int_C A\right\} = \mathbb{1} - \int_0^1 \mathrm{d}\tau\, A(\dot{Q}(\tau))$$

$$+ \int_0^1 \mathrm{d}\tau_2 \int_0^{\tau_2} \mathrm{d}\tau_1\, A(\dot{Q}(\tau_2))A(\dot{Q}(\tau_1)) + O(L^3). \tag{4.48}$$

Stokes' theorem yields immediately

$$\int_0^1 \mathrm{d}\tau\, A(\dot{Q}(\tau)) = \int_C A = \int_{S(C)} \mathrm{d}A \tag{4.49}$$

so that this term is of order L^2. The next term is conveniently decomposed into two integrals,

$$\int_0^1 \mathrm{d}\tau_2 \int_0^{\tau_2} \mathrm{d}\tau_1\, A(\dot{Q}(\tau_2))A(\dot{Q}(\tau_1)) = I_a + I_s, \tag{4.50}$$

such that the integrand of I_s is symmetric and the integrand of I_a is antisymmetric upon interchange of τ_1 and τ_2:

$$I_{s,a} = \tfrac{1}{2}\int_0^1 \mathrm{d}\tau_2 \int_0^{\tau_2} \mathrm{d}\tau_1 [A(\dot{Q}(\tau_2))A(\dot{Q}(\tau_1)) \pm A(\dot{Q}(\tau_1))A(\dot{Q}(\tau_2))]. \tag{4.51}$$

The integral I_s is easily dealt with, because due to the symmetrized integrand the noncommutativity of the matrices $A(\dot{Q}(\tau))$ is irrelevant:

$$I_s = \tfrac{1}{4} \int_0^1 d\tau_1 \, d\tau_2 [A(\dot{Q}(\tau_1)) A(\dot{Q}(\tau_2)) + A(\dot{Q}(\tau_2)) A(\dot{Q}(\tau_1))]$$

$$= \tfrac{1}{2} \left\{ \int_C A \right\}^2 = O(L^4). \tag{4.52}$$

So I_s can be neglected for our purposes. In order to calculate I_a, we introduce coordinates x^μ and write

$$A = A_\mu \, dx^\mu, \tag{4.53}$$

$$\dot{Q}(\tau) = \dot{Q}^\mu(\tau) \frac{\partial}{\partial x^\mu}\bigg|_{Q(\tau)} \tag{4.54}$$

with matrix-valued coefficient functions A_μ and $\dot{Q}^\mu = dQ^\mu/d\tau$ (see (2.12)). Within the desired accuracy we get

$$I_a = \tfrac{1}{2} \int_0^1 d\tau_2 \int_0^{\tau_2} d\tau_1 [A_\mu(Q(\tau_2)) \dot{Q}^\mu(\tau_2) A_\nu(Q(\tau_1)) \dot{Q}^\nu(\tau_1)$$

$$- A_\mu(Q(\tau_1)) \dot{Q}^\mu(\tau_1) A_\nu(Q(\tau_2)) \dot{Q}^\nu(\tau_2)]$$

$$= \tfrac{1}{2} A_\mu(Q(0)) A_\nu(Q(0)) \int_0^1 d\tau_2 \int_0^{\tau_2} d\tau_1$$

$$\times [\dot{Q}^\mu(\tau_2) \dot{Q}^\nu(\tau_1) - \dot{Q}^\mu(\tau_1) \dot{Q}^\nu(\tau_2)] + O(L^3)$$

$$= \tfrac{1}{2} A_\mu(Q(0)) A_\nu(Q(0)) \int_0^1 d\tau \left[\frac{dQ^\mu}{d\tau} Q^\nu(\tau) - \frac{dQ^\nu}{d\tau} Q^\mu(\tau) \right] + O(L^3)$$

$$= \tfrac{1}{2} A_\mu(Q(0)) A_\nu(Q(0)) \int_C (x^\nu \, dx^\mu - x^\mu \, dx^\nu) + O(L^3). \tag{4.55}$$

Applying Stokes' theorem one finds

$$I_a = \tfrac{1}{2} A_\mu(Q(0)) A_\nu(Q(0)) \int_{S(C)} d(x^\nu \, dx^\mu - x^\mu \, dx^\nu) + O(L^3)$$

$$= - \int_{S(C)} A_\mu(Q(0)) \, dx^\mu \wedge A_\nu(Q(0)) \, dx^\nu + O(L^3)$$

$$= - \int_{S(C)} A \wedge A + O(L^3). \tag{4.56}$$

So we end up with

$$P \exp\left\{ - \int_C A \right\} = \mathbb{1} - \int_{S(C)} (dA + A \wedge A) + O(L^3). \tag{4.57}$$

Comparing with (4.47), one defines the field strength (or curvature) 2-form F:

$$F := dA + A \wedge A. \tag{4.58}$$

Like A, F is $\mathfrak{g}$-valued as the following calculation shows:

$$A \wedge A = A_\mu A_\nu \, dx^\mu \wedge dx^\nu = \tfrac{1}{2}(A_\mu A_\nu \, dx^\mu \wedge dx^\nu - A_\nu A_\mu \, dx^\mu \wedge dx^\nu)$$

$$= \tfrac{1}{2}[A_\mu, A_\nu] \, dx^\mu \wedge dx^\nu = \tfrac{1}{2}[A, A] \tag{4.59}$$

(see the discussion after (1.31)). So we can write

$$F = \tfrac{1}{2} F_{\mu\nu} \, dx^\mu \wedge dx^\nu \tag{4.60}$$

with the ordinary field strength tensor

$$F_{\mu\nu} = \partial_\mu A_\nu - \partial_\nu A_\mu + [A_\mu, A_\nu]. \tag{4.61}$$

Under a gauge transformation, F transforms according to the adjoint representation; i.e. if

$$A' = \gamma A \gamma^{-1} + \gamma \, d\gamma^{-1} \tag{4.62}$$

then

$$F' = dA' + A' \wedge A' = \gamma F \gamma^{-1}. \tag{4.63}$$

4.4 The exterior covariant derivative

Having introduced a connection (gauge field) A, we are able to construct the exterior covariant derivative. We start with a 0-form Φ taking values in the vector space V, which carries a representation of G. Let C be a curve with parameter representation Q. Parallel translate $\Phi(Q(\tau))$ along an infinitesimal arc of C from $Q(\tau)$ to $Q(\tau + \varepsilon)$ with the result (see (4.37))

$$[1 - \varepsilon A(\dot{Q}(\tau))] \Phi(Q(\tau)) \tag{4.64}$$

and form the difference between the vectors $\Phi(Q(\tau + \varepsilon))$ and (4.64), both of which are to be considered as representing elements of $V_{Q(\tau + \varepsilon)}$. Note how the connection enters the construction thereby overcoming the problems mentioned after (4.35). As ε tends to zero, we get:

$$\lim_{\varepsilon \to 0} \frac{1}{\varepsilon} [\Phi(Q(\tau + \varepsilon)) - \Phi(Q(\tau)) + \varepsilon A(\dot{Q}(\tau)) \Phi(Q(\tau))]$$

$$= \frac{d}{d\tau} \Phi(Q(\tau)) + A(\dot{Q}(\tau)) \Phi(Q(\tau))$$

$$= d\Phi|_{Q(\tau)}(\dot{Q}(\tau)) + (A \wedge \Phi)|_{Q(\tau)}(\dot{Q}(\tau))$$

$$= D\Phi|_{Q(\tau)}(\dot{Q}(\tau)). \tag{4.65}$$

In the last step we have introduced the exterior covariant derivative of Φ, a 1-form with values in V:

$$D\Phi := d\Phi + A \wedge \Phi. \tag{4.66}$$

Since Φ is a 0-form, the $\wedge$ symbol could have been omitted. However, (4.66) makes sense also for V-valued p-forms Φ with $p > 0$, and this generalization will be very useful. Therefore we consider the exterior covariant derivative D as mapping V-valued p-forms into V-valued $(p + 1)$-forms.

In terms of components with respect to some basis in V, (4.66) reads:

$$(D\Phi)^i = d\Phi^i + \sum_{j=1}^{r} A^i{}_j \wedge \Phi^j, \quad i = 1, 2, \ldots, r. \tag{4.67}$$

The matrix of 1-forms $(A^i{}_j)$ is the representative of the g-valued 1-form A in the representation according to which Φ transforms. In particular, if Φ is a G-singlet, i.e. transforms according to the trivial representation of G, we have $D = d$. Note, however, that an expression like DA has no meaning, because A does not transform according to a linear (homogeneous) representation of G.

The gauge covariance of D, which is built into the above construction from the outset, is now easily confirmed. If

$$\begin{aligned} \Phi' &= \gamma\Phi, \\ A' &= \gamma A \gamma^{-1} - (d\gamma)\gamma^{-1} \end{aligned} \tag{4.68}$$

then

$$(D\Phi)' = d\Phi' + A' \wedge \Phi' = \gamma D\Phi. \tag{4.69}$$

Let Φ be a V-valued p-form and ψ a q-form which takes values in the appropriate field, $\mathbb{R}$ or $\mathbb{C}$, and transforms as a G-singlet. The Leibniz rule for d entails the following Leibniz rule for D:

$$D(\psi \wedge \Phi) = (d\psi) \wedge \Phi + (-1)^q \psi \wedge D\Phi. \tag{4.70}$$

Similar equations hold for wedge products of forms with values in other spaces.

One of the important properties of the exterior derivative is that $d^2 = 0$. The exterior covariant derivative behaves differently:

$$\begin{aligned} D^2\Phi &= d(d\Phi + A \wedge \Phi) + A \wedge (d\Phi + A \wedge \Phi) \\ &= (dA + A \wedge A) \wedge \Phi = F \wedge \Phi. \end{aligned} \tag{4.71}$$

The appearance of F in this formula, as well as in the expression (4.57) for the parallel transporter around an infinitesimal closed curve indicates that the nonvanishing of D^2 is connected with the path dependence of parallel translation.

As mentioned above, F transforms according to the adjoint representation under gauge transformations. Therefore DF is well defined:

$$DF = dF + [A, F]. \tag{4.72}$$

We calculate dF in the following manner:

$$dF = d(dA + A \wedge A) = (dA) \wedge A - A \wedge dA$$

$$= (F - A \wedge A) \wedge A - A \wedge (F - A \wedge A) = -[A, F]. \tag{4.73}$$

So we have derived the Bianchi identity

$$DF = 0, \tag{4.74}$$

which generalizes the homogeneous Maxwell equations.

4.5 Yang–Mills theories

In the general discussion up to now it was not necessary to introduce a metric. (The length in section 4.3 was used for bookkeeping purposes only.) However, if we want to write down an action for a system of gauge fields A and scalar fields Φ, we need a metric, since we have to use the Hodge star. Furthermore, we restrict ourselves to the case where our symmetry group G is compact and simple or equal to $U(1)$. In terms of forms, the Yang–Mills action reads

$$\frac{1}{g^2} \int \mathrm{tr}(F \wedge *F) + \int (D\Phi)^+ \wedge *D\Phi - m^2 \int \Phi^+ \wedge *\Phi. \tag{4.75}$$

Here g is the gauge coupling constant, m the scalar mass, Φ is considered as a column vector of complex valued 0-forms, and Φ^+ denotes the hermitian conjugate of Φ. The first term in (4.75) is the nonabelian generalization of the first term of (4.27). It is easy to verify that the action (4.75) is invariant under a local gauge transformation

$$\Phi \to \gamma\Phi, \quad A \to \gamma A \gamma^{-1} + \gamma \, d\gamma^{-1}. \tag{4.76}$$

At this place, some remarks about dimensions are useful. We have already put $c = 1$. Since we consider only classical, not quantized field theories, there is no natural unit of action. Nevertheless, we choose our units such that $\hbar = 1$. So we are left with only one basic unit, say that of length. The dimension of length is attributed to the vector fields $\partial/\partial x^\mu$. Consequently, the 1-forms dx^μ of the dual frame have dimension $(\text{length})^{-1}$. Note that the coordinates x^μ are dimensionless. The exterior derivative d, mapping p-forms into $(p+1)$-forms, carries dimension $(\text{length})^{-1}$. So A, a 1-form with Lie algebra valued, hence dimensionless, coefficient functions A_μ, acquires the dimension $(\text{length})^{-1}$, and for F we get dimension $(\text{length})^{-2}$. Since $*$ on 2-forms is dimensionless, we find that $F \wedge *F$ has dimension $(\text{length})^{-4}$. Because in the process of integration $F \wedge *F$ is evaluated on the vector fields $\partial/\partial x^\mu$ with dimension of length (see (2.87)), the integral $\int \text{tr}(F \wedge *F)$ in the action (4.75) is dimensionless. So we end up with the result that the coupling constant g of Yang–Mills theory in four dimensions is dimensionless.

In order to recast the Lagrangian into its conventional form, we choose a basis $T_a, a = 1, 2, \ldots, \dim G$, of $\mathfrak{g}$ normalized (!) according to

$$\text{tr}(T_a T_b) = -\tfrac{1}{2}\delta_{ab}. \tag{4.77}$$

Then we have

$$[T_a, T_b] = \sum_c f_{ab}{}^c T_c \tag{4.78}$$

with totally antisymmetric structure constants $f_{ab}{}^c$. Writing

$$F = \tfrac{1}{2}\sum_a F^a{}_{\mu\nu} T_a \, dx^\mu \wedge dx^\nu, \tag{4.79}$$

$$A = \sum_a A^a{}_\mu T_a \, dx^\mu \tag{4.80}$$

we get for the components of the V-valued 1-form $D\Phi$

$$(D\Phi)^i = (D_\mu\Phi)^i \, dx^\mu \tag{4.81}$$

with

$$(D_\mu\Phi)^i = \partial_\mu\Phi^i + \sum_a \sum_j A^a{}_\mu (T_a)^i{}_j \Phi^j, \tag{4.82}$$

and the Lagrangian takes the form

$$\frac{1}{g^2}\text{tr}(F \wedge *F) + (D\Phi)^+ \wedge *D\Phi - m^2\Phi^+ \wedge *\Phi$$

$$= \left\{ -\frac{1}{4g^2}\sum_a F^a{}_{\mu\nu} F^{a,\mu\nu} + \sum_i \overline{(D_\mu\Phi)^i}(D^\mu\Phi)^i \right.$$

$$\left. - m^2\sum_i \overline{\Phi^i}\Phi^i \right\} dx^0 \wedge dx^1 \wedge dx^2 \wedge dx^3. \tag{4.83}$$

The field equations follow from (4.75) by variation of A and Φ:

$$\text{tr}\left((D*F)T_a\right) = \tfrac{1}{2}g^2 *(\Phi^+ T_a D\Phi - (D\Phi)^+ T_a \Phi),$$

$$a = 1, 2, \ldots, \dim G, \tag{4.84}$$

$$*D*D\Phi - m^2\Phi = 0. \tag{4.85}$$

If there are no matter fields, we have the field equation

$$D*F = 0, \tag{4.86}$$

a nonlinear second-order partial differential equation for A. Moreover, F satisfies the Bianchi identity

$$DF = 0, \tag{4.87}$$

which is of purely geometrical origin. According to (3.18)

$$**F = -F \tag{4.88}$$

in four-dimensional Minkowski space and

$$**F = F \tag{4.89}$$

in Euclidean space. Consequently, in the latter case there exist self-dual gauge fields, i.e. fields with

$$F = *F \tag{4.90}$$

and anti-self-dual gauge fields:

$$F = -*F. \tag{4.91}$$

For (anti-) self-dual fields the field equation (4.86) is automatically satisfied due to the Bianchi identity. So one can find solutions of the second-order differential equation (4.86) by solving the first-order equation (4.90) or (4.91) for A.

4.6 Lattice gauge theory

At this point we want to make a few remarks on lattice gauge theory, because the formulation of gauge fields on a lattice is very much inspired by geometrical considerations. To put a field theory on the lattice, one replaces the continuum of space or Euclidean spacetime by a set of discrete points. In the simplest (and most often treated) case, these points form a regular (hyper) cubic lattice. A gauge field is then described by parallel transporters along the straight lines, called links, connecting nearest neighbour lattice points. So, if $l_{x,y}$ is the link pointing from x to y, the corresponding lattice gauge field variable is

$$U(l_{x,y}) := \Gamma[l_{x,y}] \in G. \qquad (4.92)$$

Local gauge transformations $\gamma(x)$ are now, of course, defined only on the lattice points and transform the link variables according to (see (4.36)):

$$U(l_{x,y}) \to \gamma(y) U(l_{x,y}) \gamma^{-1}(x). \qquad (4.93)$$

4.7 Summary

We conclude this chapter with a few summarizing remarks. The maps γ from spacetime $\mathcal{U}$ into the group G, i.e. the local gauge transformations, form an infinite-dimensional group with multiplication defined pointwise. We call it a gauge group and denote it by $^{\mathcal{U}}G$. (Caution: Many authors call G the gauge group.) The gauge field (connection) A transforms according to an affine representation of the gauge group:

$$A' = \gamma A \gamma^{-1} + \gamma \, d\gamma^{-1}, \quad \gamma \in {}^{\mathcal{U}}G. \qquad (4.94)$$

An infinitesimal gauge transformation is written as

$$\gamma = e^{\Omega} = \mathbb{1} + \Omega + \cdots, \qquad (4.95)$$

where Ω maps $\mathcal{U}$ into the Lie algebra $\mathfrak{g}$. These mappings Ω form an infinite-dimensional Lie algebra, the so-called gauge algebra $^{\mathcal{U}}\mathfrak{g}$. The behaviour of A under such an infinitesimal gauge transformation is given by

$$A' - A = -[A, \Omega] - d\Omega = -D\Omega, \qquad (4.96)$$

i.e. an affine representation of the gauge algebra. The field strength

$$F = dA + \tfrac{1}{2}[A, A] \qquad (4.97)$$

transforms according to a linear representation of the gauge group,

$$F' = \gamma F \gamma^{-1}, \qquad (4.98)$$

infinitesimally:

$$F' - F = -[F, \Omega]. \qquad (4.99)$$

For p-forms Φ taking values in a vector space which carries a linear representation of G, we have defined the exterior covariant derivative

$$D\Phi = d\Phi + A \wedge \Phi. \qquad (4.100)$$

It is such that $D\Phi$ transforms like Φ:

$$D(\gamma\Phi) = \gamma D\Phi. \qquad (4.101)$$

Furthermore, we have the Bianchi identity

$$DF = 0 \tag{4.102}$$

and the relation

$$D^2\Phi = F \wedge \Phi. \tag{4.103}$$

We call gauge theory a theory invariant under a gauge group, whereas a Yang–Mills theory is a special case of a gauge theory with an action of type (4.75) and compact symmetry group G.

Problems

4.1 Show that Maxwell's equations in terms of differential forms are given by (4.9) and (4.10).

4.2 Verify the transformation laws (4.63) and (4.69).

4.3 Prove the Leibniz rule (4.70).

4.4 Show that the action (4.75) is gauge invariant.

4.5 Derive the field equations (4.84), (4.85) from the action (4.75).

5
Einstein–Cartan theory

5.1 The equivalence principle

Up to now we have defined Yang–Mills theory in flat space. However, our formalism is valid more generally in spaces (open subsets of $\mathbb{R}^4$ so far) with arbitrary metric. This additional freedom is used to describe gravity. Indeed we make the assumption: Gravity is entirely described by a metric of Minkowski signature, $+ - - -$. We call this assumption the 'equivalence principle' although it does not correspond to Einstein's original formulation. What Einstein called the equivalence principle will in our approach follow as a corollary.

By definition, a metric is a family of bilinear forms and we do not know how to write field equations for such objects. Therefore we choose a frame $\beta^i(x), i = 0, 1, 2, 3$. With respect to this frame the metric is given by a matrix of real functions $g_{ij}(x)$ and we can write down differential equations for the β^i and g_{ij}.

Under a change of frame

$$\beta' = \gamma\beta, \quad \gamma \in {}^{\#}GL_4^+ \tag{5.1}$$

(we restrict ourselves to GL_4^+, the subgroup of GL_4 matrices with positive determinant, because we shall use the Hodge star later on), the matrix of the metric transforms as

$$g' = \gamma^{-1T}g\gamma^{-1}. \tag{5.2}$$

According to our equivalence principle, the choice of frame should be irrelevant. In other words, our task is to find differential equations for β^i and g_{ij} which are covariant under the gauge group ${}^{\#}GL_4^+$ according to its representations (5.1) and (5.2).

For this we shall use the same trick as in the last chapter: We introduce a connection and write down an invariant action. Its Euler–Lagrange equations are then the desired covariant differential equations.

Now we have an obvious problem: In the Yang–Mills case the connection represented independent physical fields, corresponding to the gauge bosons such as the photon, $W^\pm$, Z_0. Here we have just proclaimed the principle not to introduce new fields beside the metric; the solution will show up later. Meanwhile let us postulate the existence of a gl_4 connection together with its transformation law: If, with respect to the frame β^i, the connection is represented by the matrix $\Gamma^i{}_j$ of 1-forms, then with respect to β'^i it is given by the matrix

$$\Gamma' = \gamma\Gamma\gamma^{-1} + \gamma\,d\gamma^{-1}. \tag{5.3}$$

Compare with section 4.2 and note the two different meanings of the letter Γ. Because of the second, inhomogeneous term this transformation law is sometimes called 'nontensorial'. By definition gl_4 is the vector space of real 4×4 matrices, the Lie algebra of GL_4, and also of GL_4^+.

For convenience we collect the formulas for infinitesimal transformations,

$$\gamma = e^\Omega = \mathbb{1} + \Omega + O(\Omega^2), \tag{5.4}$$

where Ω is a gl_4-valued 0-form:

$$\Omega \in {}^{\#}gl_4, \tag{5.5}$$

$$\beta' - \beta = \Omega\beta, \tag{5.6}$$

$$g' - g = -\Omega^T g - g\Omega, \tag{5.7}$$

$$\Gamma' - \Gamma = -d\Omega - [\Gamma, \Omega] = -D\Omega. \tag{5.8}$$

5.2 Cartan's structure equations

As before we define the field strength as a 2-form with values in gl_4:

$$R := d\Gamma + \tfrac{1}{2}[\Gamma, \Gamma], \tag{5.9}$$

usually called curvature in this context. R transforms homogeneously:

$$R' = \gamma R\gamma^{-1}, \tag{5.10}$$

infinitesimally:

$$R' - R = [\Omega, R]. \tag{5.11}$$

Similarly we define torsion, a 2-form with values in $\mathbb{R}^4$:

$$T := d\beta + \Gamma \wedge \beta = D\beta, \tag{5.12}$$

which transforms as

$$T' = \gamma T, \tag{5.13}$$

infinitesimally:

$$T' - T = \Omega T. \tag{5.14}$$

The definitions of curvature and torsion are called Cartan's second and first structure equation.

The two Bianchi identities follow immediately:

$$DR = dR + [\Gamma, R] = 0, \tag{5.15}$$

$$DT = DD\beta = R \wedge \beta. \tag{5.16}$$

5.3 Metric connections

In the last chapter we have seen that a connection determines the parallel transport of vectors carrying a linear representation of the underlying group. This is now GL_4^+. At any point $x \in \mathcal{U}$ the tangent vectors v described by their components v^i with respect to a frame $b_0, \ldots, b_3$, to which $\beta^0, \ldots, \beta^3$ is dual

$$v^i = \beta^i(v), \tag{5.17}$$

$$v = v^i b_i, \tag{5.18}$$

carry the fundamental representation of GL_4^+. The transformation law follows from (5.1):

$$v'^i(x) = \gamma(x)^i{}_j v^j(x). \tag{5.19}$$

Accordingly, parallel transport of tangent vectors is well defined: Let C be a curve in $\mathcal{U}$ starting at x_0 and let

$$Q: [0, 1] \to \mathcal{U}$$

$$\tau \mapsto Q(\tau)$$

be any parametrization of C. The parallel transport of a tangent vector $v_0 \in T_{x_0}\mathcal{U}$ along C by means of a connection Γ is the one-parameter family of tangent vectors

$$v_\tau \in T_{Q(\tau)}\mathcal{U} \tag{5.20}$$

solving the differential equation (cf. (4.38))

$$\frac{d}{d\tau}v_\tau{}^i = -\Gamma^i{}_j|_{Q(\tau)}(\dot{Q}(\tau))v_\tau{}^j \tag{5.21}$$

with initial value $v_0{}^i$.

We can now ask whether parallel transport of tangent vectors preserves the metric, i.e. whether

$$g_{x_0}(v_0, w_0) = g_{Q(\tau)}(v_\tau, w_\tau). \tag{5.22}$$

If so for any curve C and all tangent vectors v_0 and w_0, the connection Γ is called metric. In other words, Γ is metric if its parallel transporters are isometries. One easily verifies that this is equivalent to

$$Dg = dg - \Gamma^T g - g\Gamma = 0. \tag{5.23}$$

We call this equation the metric condition. In components it reads:

$$dg_{ij} - \Gamma^k{}_i g_{kj} - g_{ik} \Gamma^k{}_j = 0. \tag{5.24}$$

Γ is a 1-form with values in gl_4 and therefore given by $4 \times 16 = 64$ real functions

$$\Gamma^i{}_{jk} := \Gamma^i{}_j(b_k),$$

which we call unknowns because according to the equivalence principle they have to be eliminated.

Now the metric condition is a system of equations for 1-forms, i.e. $4 \times 10 = 40$ real equations, linear in the 64 unknowns $\Gamma^i{}_{jk}$, and the metric condition is a first step towards the equivalence principle.

This becomes more transparent if we choose an orthonormal frame e^i. We also reserve a special letter ω for the connection if it is expressed with respect to an orthonormal frame and if it is metric. The connection ω is sometimes called the spin connection. In an orthonormal frame the matrix $g_{ij} = \eta_{ij}$ is constant and the metric condition reduces to

$$\omega^T \eta + \eta \omega = 0, \tag{5.25}$$

which just means that ω is a 1-form with values in the Lorentz algebra

$$\omega \in \Lambda^1(\mathcal{U}, so(1,3)).$$

This was of course to be expected because by definition the Lorentz group preserves the metric. We recall that this reduction of the symmetry group from GL_4^+ to $SO(1,3)$ is just the orthonormalization procedure of Gram and Schmidt.

Let us count degrees of freedom (real functions): We start with a frame β. Consisting of four 1-forms, it is represented by $4 \times 4 = 16$ functions. The metric is now described by a symmetric matrix of 0-forms (10 functions). We postulate covariance under GL_4^+, a 16-dimensional group, and thereby subtract 16 degrees of freedom. In addition to the frame and the metric we had to introduce a gl_4 connection Γ (64 functions). The metric condition gives 40 linear covariant constraints on the connection.

In an orthonormal frame the count is as follows: The orthonormal frame contains four 1-forms (16 functions). The metric now represented by the constant matrix η contains no function. The invariance group is reduced to $SO(1,3)$, a six-dimensional Lie group. In addition we need the $so(1,3)$ connection ω with $4 \times 6 = 24$ degrees of freedom:

β	g	$\gamma \in GL_4^+$	Γ	metr. cond.
16	10	-16	64	-40
e	η	$\Lambda \in SO(1,3)$	ω	
16	0	-6	24	

5.4 Action and field equations

In pure Yang–Mills theory the connection is the only independent field, and the simplest gauge invariant action is quadratic in the field strength:

$$S_{\mathrm{YM}} := \frac{1}{g^2} \int_{\mathcal{U}} F^a{}_b \wedge *F^b{}_a. \tag{5.26}$$

Here we have at our disposal the homogeneously transforming frame β in addition to the connection Γ. Therefore there is already a gauge-invariant action linear in the curvature:

$$S_{\mathrm{EH}} := \frac{-1}{32\pi G} \int_{\mathcal{U}} R^a{}_b \wedge *(\beta^b \wedge \beta_a), \tag{5.27}$$

the Einstein–Hilbert action. We lower (raise) indices by means of the matrix of the metric (its inverse) with respect to the particular frame chosen (cf. the canonical isomorphism, equation (3.14)).

Let us deduce the units of the gravitational coupling constant G: If we measure tangent vectors in units of length and 1-forms in units of $(\text{length})^{-1}$, then vector components are dimensionless and the metric components $g_{a'a}$ used to lower the index of the second β have units $(\text{length})^2$. Therefore in order to have a dimensionless action, the gravitational coupling constant G must have units $(\text{length})^2$. This is one major difficulty in quantizing gravity.

Next we want to derive the field equations by varying the fields. For this we shall work in an orthonormal frame with independent fields e and ω.

This has two advantages: The Hodge star becomes more convenient, and the metric condition being already eliminated, there are no constraints to worry about. In an orthonormal frame the total action reads:

$$\frac{-1}{32\pi G}\int R^{ab}\wedge e^c\wedge e^d\varepsilon_{abcd}+\int\mathscr{L}_{\text{M}},\tag{5.28}$$

where the matter action $\int\mathscr{L}_{\text{M}}$ is a functional of e, ω, and the matter fields.

Varying e

We replace e by $e+f$ in the action and calculate the term linear in f:

$$S_{\text{EH}}[e+f,\omega]-S_{\text{EH}}[e,\omega]=\frac{-2}{32\pi G}\int f^c\wedge R^{ab}\wedge e^d\varepsilon_{abcd}+O(f^2),\tag{5.29}$$

$$\mathscr{L}_{\text{M}}[e+f]-\mathscr{L}_{\text{M}}[e]=-f^c\wedge\tau_c+O(f^2).\tag{5.30}$$

Here, τ_c is a vector-valued 3-form, the 'energy–momentum current of matter'. It is the analogue of the electromagnetic current j. Integrated over a three-dimensional spacelike volume it yields the energy–momentum of matter included in this volume. The requirement of vanishing variation leads to the Einstein equation

$$R^{ab}\wedge e^d\varepsilon_{abcd}=-16\pi G\tau_c.\tag{5.31}$$

To write it in components we put

$$R^{ab}=\tfrac{1}{2}R^{ab}{}_{rs}e^r\wedge e^s,\tag{5.32}$$

$$*\tau_c=\tau_{ca}e^a,\tag{5.33}$$

or equivalently

$$\tau_c=\tfrac{1}{6}\tau_c{}^a\varepsilon_{arsd}e^r\wedge e^s\wedge e^d.\tag{5.34}$$

With these definitions Einstein's equation becomes

$$R^{ab}{}_{rs}\varepsilon_{abcd}e^r\wedge e^s\wedge e^d=-\tfrac{32}{6}\pi G\tau_c{}^a\varepsilon_{arsd}e^r\wedge e^s\wedge e^d,\tag{5.35}$$

and applying the Hodge star one finds

$$R^{ab}{}_{rs}\varepsilon_{abcd}\varepsilon^{krsd}=-\tfrac{32}{6}\pi G\tau_c{}^a\varepsilon_{arsd}\varepsilon^{krsd}.\tag{5.36}$$

We use

$$\varepsilon^{0123}=-1,\tag{5.37}$$

$$\varepsilon^{aijk}\varepsilon_{bijk}=-6\delta^a{}_b,\tag{5.38}$$

$$\varepsilon^{abci}\varepsilon_{rsti} = -[\delta^a{}_r\delta^b{}_s\delta^c{}_t + \delta^b{}_r\delta^c{}_s\delta^a{}_t + \delta^c{}_r\delta^a{}_s\delta^b{}_t$$
$$- \delta^b{}_r\delta^a{}_s\delta^c{}_t - \delta^c{}_r\delta^b{}_s\delta^a{}_t - \delta^a{}_r\delta^c{}_s\delta^b{}_t]$$
(5.39)

and get Einstein's equation in components:

$$G_{kc} = 8\pi G \tau_{ck}$$
(5.40)

with the abbreviations

$$G^k{}_s := R^{ak}{}_{ac} - \tfrac{1}{2}R^{ab}{}_{ab}\delta^k{}_c, \quad \text{Einstein tensor,}$$
(5.41)

$$R^{ak}{}_{ac}, \quad \text{Ricci tensor,}$$

$$R^{ab}{}_{ab}, \quad \text{curvature scalar.}$$

The Einstein equations are nonlinear first-order partial differential equations for the connection components. They are also linear equations for the curvature components: 'Energy–momentum is the source of curvature'. Although the relation between energy–momentum and curvature is algebraic, curvature 'propagates' in four (and higher) dimensions: Vanishing energy–momentum does not imply zero curvature, the Schwarzschild solution and gravitational radiation being prominent examples. The reason is the following: $R^{ab}{}_{cd}$ as components of a 2-form is antisymmetric in c and d, being $so(1, 3)$-valued it is also antisymmetric in a and b. Therefore in four dimensions it has $6 \times 6 = 36$ independent components. There are, however, only $4 \times 4 = 16$ linear equations, an underdetermined system.

Varying ω

We replace ω by $\omega + \chi$ in the curvature:

$$R(\omega + \chi) = d\omega + d\chi + (\omega + \chi) \wedge (\omega + \chi) = R(\omega) + D\chi + O(\chi^2).$$
(5.42)

Note that χ is the difference of two connections and transforms homogeneously. Therefore its covariant derivative is defined and we get

$$S_{\text{EH}}[\omega + \chi] - S_{\text{EH}}[\omega] = \frac{-1}{32\pi G} \int (D\chi)^{ab} \wedge e^c \wedge e^d \varepsilon_{abcd} + O(\chi^2). \quad (5.43)$$

We integrate by parts to shift the covariant derivative away from χ:

$$0 = \int d(\chi^{ab} \wedge e^c \wedge e^d \varepsilon_{abcd})$$

$$= \int D(\chi^{ab} \wedge e^c \wedge e^d \varepsilon_{abcd})$$

$$= \int (D\chi)^{ab} \wedge e^c \wedge e^d \varepsilon_{abcd} - 2 \int \chi^{ab} \wedge (De)^c \wedge e^d \varepsilon_{abcd}. \quad (5.44)$$

The first equality follows from Stokes' theorem neglecting the boundary term, the second holds because $\chi^{ab} \wedge e^c \wedge e^d \varepsilon_{abcd}$ is Lorentz-invariant, the third is a Leibniz rule for the covariant derivative where we have used

$$(D\varepsilon)_{abcd} = d\varepsilon_{abcd} = 0. \tag{5.45}$$

This is true because the ε-symbol is invariant under $SO(1,3)$. So we end up with

$$S_{EH}[\omega + \chi] - S_{EH}[\omega] = \frac{-1}{16\pi G} \int \chi^{ab} \wedge T^c \wedge e^d \varepsilon_{abcd} + O(\chi^2). \tag{5.46}$$

We have to give the variation of the matter a Lagrangian name:

$$\mathscr{L}_M[\omega + \chi] - \mathscr{L}_M[\omega] = -\tfrac{1}{2}\chi^{ab} \wedge S_{ab} + O(\chi^2). \tag{5.47}$$

Like τ_c, the S_{ab} are 3-forms, 'the spin currents' of matter.

The field equation now reads

$$T^c \wedge e^d \varepsilon_{abcd} = -8\pi G S_{ab}. \tag{5.48}$$

This is a system of first-order partial differential equations for e. They are also linear equations for the torsion: 'Spin is the source of torsion'. Contrary to curvature, torsion does not propagate (in any dimension): Torsion is a vector-valued 2-form and has 6×4 components. On the other hand, the field equation for torsion in four dimensions consists of 6 equations for 1-forms (4×6 linear, independent equations). Therefore torsion always vanishes outside matter. Many matter Lagrangians do not depend on the connection, for example the Lagrangian describing scalar or Yang–Mills fields. Then spin current and torsion vanish identically.

We close this section with two remarks: In terms of the curvature scalar $R^{ab}{}_{ab}$, the Einstein–Hilbert action reads

$$S_{EH} = \frac{-1}{16\pi G} \int d^4x |g|^{1/2} R^{ab}{}_{ab}. \tag{5.49}$$

The field equations which we have derived in an orthonormal frame are covariant and therefore valid in any frame.

5.5 A farewell to Γ

We now come to the promised elimination of the connection in order to comply with the equivalence principle. There are two possible routes:

Einstein's view point

Einstein adds to the metric condition a second covariant constraint:

$$0 = T = d\beta + \Gamma \wedge \beta, \tag{5.50}$$

which constitutes four equations for 2-forms, i.e. $6 \times 4 = 24$ linear equations in the 64 unknowns $\Gamma^i{}_{jk}$. Together with the metric condition (5.23) we therefore have $40 + 24 = 64$ linear equations in 64 unknowns. The system admits a unique solution:

$$\Gamma^i{}_{jk} = \tfrac{1}{2}[C^i{}_{jk} - g^{ii'}g_{jj'}C^{j'}{}_{i'k} - g^{ii'}g_{kk'}C^{k'}{}_{i'j}]$$
$$+ \tfrac{1}{2}g^{ii'}[(dg_{ji'})(b_k) + (dg_{i'k})(b_j) - (dg_{kj})(b_{i'})], \tag{5.51}$$

where we have put

$$d\beta^i = \tfrac{1}{2}C^i{}_{jk}\beta^j \wedge \beta^k. \tag{5.52}$$

This metric torsion-free connection is called the 'Riemannian' or 'Levi-Civita' connection of the metric g.

In an orthonormal frame $\beta^a = e^a$ the components of the metric are constant and the second part of equation (5.51) vanishes.

$$0 = T = ddx^\mu + \Gamma^\mu{}_{\rho\sigma}dx^\rho \wedge dx^\sigma$$

implies

$$\Gamma^\mu{}_{\rho\sigma} = \Gamma^\mu{}_{\sigma\rho}. \tag{5.53}$$

We call this a bastard symmetry because it mixes form and value indices.

In a holonomic frame $\beta^\mu = dx^\mu$ the Cs are all zero and the $\Gamma^\mu{}_{\rho\sigma}$ are called Christoffel symbols. Furthermore,

The formula for the Riemannian connection is valid in all dimensions larger than one and for any signature. As an illustration we apply it to our two-dimensional surface, a piece of the unit sphere, parametrized by

$$x^1 = \varphi, \quad x^2 = \theta. \tag{5.54}$$

Here $dx^1 = d\varphi$, $dx^2 = d\theta$ is a holonomic frame with metric components

$$(g^{ij}) = \begin{pmatrix} \sin^{-2}\theta & 0 \\ 0 & 1 \end{pmatrix}, \tag{5.55}$$

and $e^1 = \sin\theta\, d\varphi, e^2 = d\theta$ is an orthonormal frame. We compute the $C^i{}_{jk}$:

$$de^1 = \cos\theta\, d\theta \wedge d\varphi = -\frac{\cos\theta}{\sin\theta} e^1 \wedge e^2, \tag{5.56}$$

$$de^2 = 0.$$

Therefore

$$C^1_{12} = -C^1_{21} = -\frac{\cos\theta}{\sin\theta}, \qquad (5.57)$$

and the other C^i_{jk} vanish. Next we calculate the Riemannian connection with respect to the orthonormal frame:

$$\omega^i_{jk} = \tfrac{1}{2}(C^i_{jk} - C^j_{ik} - G^k_{ij}), \qquad (5.58)$$

$$\omega^1_{21} = \frac{\cos\theta}{\sin\theta}, \quad \omega^1_{22} = 0,$$

$$\omega^1_2 = \frac{\cos\theta}{\sin\theta} e^1 = \cos\theta \, d\varphi. \qquad (5.59)$$

Remembering that $SO(2,0)$ is one-dimensional and abelian we find

$$R^1_2 = d\omega^1_2 = -\sin\theta \, d\theta \wedge d\varphi = e^1 \wedge e^2. \qquad (5.60)$$

Consequently, one gets

$$R^1_{212} = -R^1_{221} = -R^2_{112} = R^2_{121} = 1 \qquad (5.61)$$

and finally the curvature scalar

$$R^{ab}_{ab} = 2. \qquad (5.62)$$

We can now be sure that this two-dimensional surface is not flat; i.e. it is impossible to find a holonomic orthonormal frame because of the following theorem valid in all dimensions larger than one and for any signature.

THEOREM

The curvature 2-form of the Riemannian connection vanishes identically if and only if there exists locally (in an open neighbourhood of any point) a holonomic orthonormal frame.

Einstein considers his action, equation (5.27), as a functional of the metric only, with R the curvature of the Riemannian connection. Variation with respect to the metric then leads immediately to the Einstein equation for the metric, a system of nonlinear second-order partial differential equations. This is called second-order formalism. Of course one gets the same result by eliminating the connection with equation (5.51) from Einstein's equation derived previously in the 'first-order formalism'.

Cartan's viewpoint

Here despite the equivalence principle the connection is kept as an independent variable in addition to the metric. Only the field equations obtained by varying the connection fix the torsion as a function of the spin current. We shall see later that Lorentz-invariant spin $\frac{1}{2}$ and $\frac{3}{2}$ matter actions can only be constructed by means of the spin connection ω. In presence of such matter, spin and consequently torsion are non vanishing. Therefore, in general this so-called Einstein–Cartan theory is different from Einstein's general relativity. However, torsion does not propagate, and in the vacuum both theories are identical. In particular, Einstein–Cartan theory satisfies the equivalence principle in the vacuum. On the other hand, the spin density in the universe is small and its coupling to torsion uniquely fixed by the theory to be the gravitational constant G. As a consequence both theories are presently indistinguishable experimentally.

A few final remarks:

The field equations for torsion are again 24 linear equations in the 24 spin connection components $\omega^a{}_{bc}$ and as before in the torsionless case they can be solved to give the spin connection as a function of the frame derivatives $C^a{}_{bc}$ and the spin current.

The Einstein–Hilbert action is the only action which leads to vanishing torsion in vacuum upon variation of the connection (Yates 1980), unique of course only up to terms not involving the connection, e.g. a cosmological term

$$\frac{\Lambda}{4!} \int_{\mathscr{U}} e^a \wedge e^b \wedge e^c \wedge e^d \varepsilon_{abcd}. \tag{5.63}$$

A particular Einstein–Cartan theory with the real spin $\frac{3}{2}$ Rarita–Schwinger action for matter has in addition to the diffeomorphism (cf. chapter 6) and Lorentz gauge invariance a most remarkable symmetry mixing the spin 2 graviton described by the orthonormal frame and the spin $\frac{3}{2}$ field (Deser & Zumino 1976; Freedman, van Nieuwenhuizen & Ferrara 1976). The latter is named gravitino, and the theory $N = 1$ (one gravitino) supergravity without matter, because the gravitino is considered as gravitational radiation together with the graviton.

5.6 Energy–momentum of matter

In this section we discuss three general properties of the energy–momentum tensor τ_{ck} – symmetry, covariant conservation, positivity of τ_{00} – and calculate it for the Maxwell action.

The energy–momentum tensor can be defined in several ways. In Minkowski space, which is a vector space, translations can be defined and are used to define energy–momentum. Invariance of the action under translations then implies, via Emmy Noether's theorem, conservation of energy–momentum. For open subsets of $\mathbb{R}^4$ there are in general no translations. We defined energy–momentum by varying the metric, which enters the matter action through the Hodge star. If the metric is described by its components in an arbitrary frame, one has to vary these components, and the energy–momentum tensor is automatically symmetric. We have chosen to describe the metric by an orthonormal frame which we therefore had to vary, equation (5.30).

Symmetry

The curvature components $R^{ab}_{\ \ cd}$ are antisymmetric in ab and cd. If torsion is zero, there is a third, cyclic symmetry, a bastard symmetry:

$$R^a_{\ bcd} + R^a_{\ cdb} + R^a_{\ dbc} = 0. \tag{5.64}$$

It is easily derived by applying the Hodge star to the second Bianchi identity (5.16).

Together with the other two symmetries it implies that the Einstein tensor is symmetric:

$$G_{ck} = G_{kc}. \tag{5.65}$$

Einstein's equation then tells us that we can consistently couple only matter with symmetric energy–momentum tensors. Since we have assumed vanishing torsion, this matter cannot depend on the connection ω.

We now show that Lorentz invariance of the matter action is sufficient to guarantee this symmetry as long as the matter fields satisfy their proper field equations, i.e. as long as they are 'on shell'. We recall that Lorentz invariance follows from the description of the metric by means of an orthonormal basis.

To calculate the variation of the matter Lagrangian under an infinitesimal Lorentz transformation $(\Omega^a_{\ b}) \in so(1,3)$ we can immediately use the defining equation of energy–momentum (5.30) with

$$f^a = -\Omega^a_{\ b} e^b, \tag{5.66}$$

because on shell the variation of the matter fields does not contribute:

$$0 = \Omega^c_{\ b} e^b \wedge \tfrac{1}{6} \tau_c^{\ a} \varepsilon_{arsd} e^r \wedge e^s \wedge e^d. \tag{5.67}$$

Applying the Hodge star we find:

$$0 = \Omega^c{}_b \tfrac{1}{6} \tau_c{}^a \varepsilon_{arsd} \varepsilon^{brsd} = -\Omega^{cb} \tau_{cb}. \tag{5.68}$$

Ω^{cb} being an arbitrary antisymmetric matrix, this equation implies the bastard symmetry

$$\tau_{cb} = \tau_{bc}. \tag{5.69}$$

In the case of nonvanishing torsion both the Einstein and the energy–momentum tensor have an antisymmetric part. A similar calculation shows that again Lorentz invariance of the matter action is sufficient for consistency. A major motivation to consider gravity theories with nonvanishing torsion was the remark that in classical mechanics spin leads to a nonsymmetric stress tensor (Papapetrou 1949).

Covariant conservation

Let us take the covariant derivative of Einstein's equation (5.31). If torsion is zero, the first Bianchi identity implies that the covariant derivative of the lhs is zero and consistently the energy–momentum of matter must be (covariantly) conserved:

$$D\tau = 0. \tag{5.70}$$

This is just as in the Maxwell case where only conserved currents can be coupled to the electromagnetic field.

Invariance of the matter action under the group of diffeomorphisms is sufficient to guarantee this conservation law on shell if torsion vanishes. For the proof we use again equation (5.30) and put

$$f^a = L_v e^a, \tag{5.71}$$

where L_v is the Lie derivative representing infinitesimal diffeomorphisms. It will be defined in the next chapter.

If torsion is nonzero, neither side of Einstein's equation is conserved.

Positivity

An additional physical assumption, which is not a consistency requirement, is the positivity of the energy density τ_{00} in all orthonormal frames. In the following we briefly sketch one of its implications. Up to now we have discussed the energy–momentum density of matter. For the gravitational part there is no such object. The variation of the Einstein–Hilbert action with respect to the orthonormal frame vanishes because of Einstein's equation. However, for asymptotically flat spacetime with no torsion there is a definition of total energy of gravity and matter (Einstein 1916; 1918; Arnowitt, Deser & Misner 1960 a, b; 1961) which has the following propitious properties:

(a) it is conserved, i.e. time independent;
(b) the total energy of two isolated systems is the sum of the two energies;
(c) it is positive under the assumption of positive energy density of matter, and zero only for matter-free, flat space, which is therefore the stable ground state.

The positivity proof is difficult and was completed only in 1979 (Schoen & Yan 1979).

Energy–momentum of the Maxwell action

We calculate the energy–momentum of the electromagnetic field. The calculation for a general Yang–Mills field is analogous. The Maxwell Lagrangian

$$\mathscr{L}_M = \frac{1}{2g^2} F \wedge *F \tag{5.72}$$

depends on the orthonormal frame only via the Hodge star. In the following we indicate by $*|_e$ the Hodge star for the metric described by the orthonormal frame e. Its dependence on e is implicit and the variation not straightforward. We start from an identity which is easy to verify:

$$e^a \wedge e^b \wedge *|_e F = \tfrac{1}{2} \varepsilon^{ab}{}_{cd} e^c \wedge e^d \wedge F. \tag{5.73}$$

Now replacing e by $e + f$ and neglecting terms quadratic in f we get:

$$f^a \wedge e^b \wedge *|_e F + e^a \wedge f^b \wedge *|_e F$$
$$+ e^a \wedge e^b \wedge (*|_{e+f} F - *|_e F) = \varepsilon^{ab}{}_{cd} f^c \wedge e^d \wedge F. \tag{5.74}$$

We multiply by $(1/4g^2) F_{ab}$ and contract:

$$\mathscr{L}_M[e+f] - \mathscr{L}_M[e] = \frac{1}{g^2} f^c (\tfrac{1}{4} F^{ab} \varepsilon_{abcd} e^d \wedge F - \tfrac{1}{2} F_{cb} e^b \wedge *|_e F). \tag{5.75}$$

Therefore

$$\tau_c = \frac{-1}{8g^2} F^{ab} F_{rs} \varepsilon_{abcd} e^d \wedge e^r \wedge e^s$$
$$+ \frac{1}{8g^2} F_{cb} F^{ad} \varepsilon_{adrs} e^b \wedge e^r \wedge e^s \tag{5.76}$$

and

$$*\tau_c = \frac{-1}{g^2}(\tfrac{1}{4}F^{ab}F_{ab}\eta_{kc} + F_{ca}F^a{}_k)e^k = \tau_{ck}e^k, \tag{5.77}$$

where we have used

$$\varepsilon_{adrs}\varepsilon^{bkrs} = -2(\delta^b{}_a\delta^k{}_d - \delta^k{}_a\delta^b{}_d). \tag{5.78}$$

Recalling

$$(F^{ab}) = -\,\mathrm{i}g \begin{pmatrix} 0 & E_1 & E_2 & E_3 \\ -E_1 & 0 & B_3 & -B_2 \\ -E_2 & -B_3 & 0 & B_1 \\ -E_3 & B_2 & -B_1 & 0 \end{pmatrix} \tag{5.79}$$

we get

$$\tau_{00} = \tfrac{1}{2}(E^2 + B^2). \tag{5.80}$$

Finally we remark that this definition of energy–momentum is Yang–Mills gauge invariant, because it is the variation of the gauge invariant Lagrangian with respect to the Yang–Mills gauge invariant frame.

5.7 The Einstein gauge

From Einstein's formulation it is difficult to see that general relativity is in fact a gauge theory. This is the case because Einstein fixed the gauge from the start by choosing a coordinate system and then worked in its holonomic frame. Usually gauge fixing, while helpful in concrete calculations, masks the general structure of the theory.

The Einstein gauge is particularly economical since in many situations a coordinate system is needed anyhow. Geodesics, a central theme in general relativity, are one example to be discussed below. On the other hand Einstein's gauge has two major shortcomings:

(a) It is a GL_4^+ gauge. We shall see later that GL_4^+ does not admit spinor representations. They can be introduced only after reduction of GL_4^+ to $SO(1,3)$ by taking orthonormal frames incompatible with Einstein's gauge.

(b) The Einstein gauge does not completely fix the gauge. However, the remaining freedom, the coordinate transformations, do not form a group because of possible coordinate singularities.

A gauge fixing is achieved by choosing 'harmonic coordinates'. These are coordinates x^μ such that

$$\delta\,\mathrm{d}x^\mu = -\,\square\,x^\mu = 0. \tag{5.81}$$

For example, with respect to the Euclidean metric Cartesian coordinates are harmonic; polar coordinates are not.

We collect a few abuses of language that are particularly frequent in the context of Einstein's gauge. Coordinate transformations are not diffeomorphisms. The latter are globally defined and form a nonabelian group. The former are in general not globally defined – this will be crucial in the context of manifolds – and do not form a group. The diffeomorphism group is not a gauge group and Γ is not its connection.

The change of frame from a holonomic frame dx^μ to an orthonormal frame e^a is given by an element γ of the GL_4^+ gauge group:

$$e^a = \gamma^a{}_\mu dx^\mu, \quad \gamma \in {}^{\mathscr{U}}GL_4^+. \tag{5.82}$$

The connection components with respect to a holonomic frame, Γ, and with respect to an orthonormal frame, ω, are related by equation (5.3)

$$\omega = \gamma \Gamma \gamma^{-1} + \gamma \, d\gamma^{-1}, \tag{5.83}$$

which in holonomic components reads:

$$\partial_\nu \gamma^a{}_\mu - \gamma^a{}_\alpha \Gamma^\alpha{}_{\mu\nu} + \omega^a{}_{b\nu} \gamma^b{}_\mu = 0. \tag{5.84}$$

The ${}^{\mathscr{U}}GL_4^+$-element $(\gamma^a{}_\mu)$ is often denoted $(e^a{}_\mu)$ and also called the vierbein. The last equation is then written

$$\mathbf{D}_\nu e^a{}_\mu = 0$$

and confused with the metric condition

$$\mathbf{D}_\nu g_{\alpha\mu} = 0 \tag{5.85}$$

by calling the vierbein $e^a{}_\mu$ a square root of the metric, cf. equation (3.8).

5.8 Geodesics

DEFINITION

A parameterized curve $Q(\tau)$ is called geodesic with initial point $x_0 \in \mathscr{U}$ and in direction of $v_0 \in T_{x_0}\mathscr{U}$ if

$$Q(0) = x_0, \tag{5.86}$$

if its tangent vector at x_0 is v_0; in coordinates:

$$\frac{dQ^\mu(0)}{d\tau} = dx^\mu(v_0), \tag{5.87}$$

and if its tangent vectors for later τ result from v_0 by parallel transport; in coordinates:

$$\frac{d^2 Q^\mu}{d\tau^2} + \Gamma^\mu_{\;\nu\alpha} \frac{dQ^\nu}{d\tau} \frac{dQ^\alpha}{d\tau} = 0. \tag{5.88}$$

This is a nonlinear system of four second-order ordinary differential equations in four unknowns $Q^\mu(\tau)$ with initial conditions $Q^\mu(0)$ and $dQ^\mu/d\tau(0)$ and there always exists locally (for small enough τ) a unique solution. Note that whilst only the point set of a curve enters the definition of parallel transport, a geodesic is a curve with a specific parameterization $Q(\tau)$.

The name geodesic is justified by the following theorem.

THEOREM

Let g be a positive-definite metric. Geodesics with respect to the metric, torsion-free connection extremize locally the length of a curve

$$\int_0^1 d\tau \left[g_{\mu\nu}(Q(\tau)) \frac{dQ^\mu}{d\tau}(\tau) \frac{dQ^\nu}{d\tau}(\tau) \right]^{1/2}. \tag{5.89}$$

Let us consider again a piece of the two-dimensional unit sphere. We have already calculated its Riemannian connection. It follows that geodesics are just segments of a big circle, obtained for example by stretching a rubber band over a globe. The parallel transport of a vector is particularly simple along geodesics. By the metric condition, we have to keep a constant angle with the geodesic.

In flat space one can always find coordinates such that the connection expressed in their holonomic frame is zero, e.g. Cartesian coordinates for the Euclidean metric. Geodesics in flat space are therefore straight lines. Conversely, for any space with connection we can construct local coordinates x^μ around a fixed point $x_0 \in \mathcal{U}$ such that all curves through x_0 described by straight lines in these coordinates,

$$Q^\mu(\tau) = \tau a^\mu, \quad a^\mu \in \mathbb{R}, \tag{5.90}$$

are geodesics. These coordinates are called geodesic, normal or Gaussian. Their construction is as follows. Let x^μ be Cartesian coordinates of the tangent space at x_0 defined by four vectors orthonormal with respect to g_{x_0}. For any small enough vector in $T_{x_0}\mathcal{U}$ described by the four numbers x^μ we construct the geodesic $Q(\tau)$ with initial point x_0 and in the direction of this vector. The point $Q(1)$ in $\mathcal{U}$ has then by definition the coordinates x^μ.

With respect to geodesic coordinates the metric components at x_0 are Minkowskian:

$$g_{\mu\nu}(x_0) = \eta_{\mu\nu}, \qquad (5.91)$$

and the symmetric part of the connection components vanishes at x_0:

$$\Gamma^\mu{}_{\nu\alpha}(x_0) + \Gamma^\mu{}_{\alpha\nu}(x_0) = 0. \qquad (5.92)$$

The geodesic coordinates are the coordinates of Einstein's freely falling lift and they express his principle of equivalence.

The equivalence principle is independent of the choice of the action. Nevertheless Einstein's equation implies that a spinless test mass in torsion-free space flies on geodesics. The proof given by Einstein, Infeld & Hoffmann (1938) and subsequently by Einstein & Infeld (1940, 1949) is difficult: Einstein's equations, being nonlinear, cannot be viewed as differential equations for distributions and point masses described by delta-like energy densities make no sense. There is, however, a simple argument which plausibly explains why trajectories of test particles are determined by the field equations in Einstein's theory, in contrast to Maxwell's theory. Forces and trajectories follow from energy–momentum conservation of matter and we have seen that Einstein's equations imply energy–momentum conservation, while Maxwell's equations only imply charge conservation. Energy–momentum of charged matter has to be defined independently by the Lorentz force law.

5.9 Geometric interpretation of curvature and torsion

In this section we do not need a metric. We start from a frame β and a connection, which with respect to the frame β we denote by Γ. Curvature and torsion are defined by Cartan's structure equations (5.9) and (5.12). They are both 2-forms to be evaluated on two tangent vectors. Let us fix a point x_0 in $\mathcal{U}$ and consider two vectors v and w at x_0,

$$v, w \in T_{x_0}\mathcal{U}.$$

We shall now construct a family of 'geodesic parallelograms' from v and w indexed by τ_0, see fig. 5.1. The construction works for small enough τ_0 (locally). Let $Q_1(\tau)$ be the geodesic starting at x_0 in direction of v. Parallel transport w from x_0 to $Q_1(\tau_0)$ along $Q_1(\tau)$. Call the resultant vector w_1. Construct the geodesic $Q_2(\tau)$ starting at $Q_1(\tau_0)$ in direction of w_1. Let v_2 be the parallel transport of v from x_0 to $Q_2(\tau_0)$ along Q_1 and Q_2. Construct the geodesic $Q_3(\tau)$ starting at $Q_2(\tau_0)$ in direction $-v_2$. Then w_3 is the parallel transport of w from x_0 to $Q_3(\tau_0)$ along Q_1, Q_2, and Q_3,

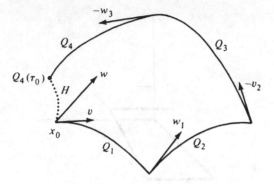

Fig. 5.1. A geodesic 'parallelogram'.

and $Q_4(\tau)$ is the geodesic from $Q_3(\tau_0)$ in direction $-w_3$. This geodesic ends at $Q_4(\tau_0)$, a point in $\mathscr{U}$. Considered as a function of τ_0 the point $Q_4(\tau_0)$ sweeps out a curve $H(\tau_0)$ with initial point

$$H(0) = x_0. \tag{5.93}$$

This curve has the following properties: Its tangent vector in x_0 is zero,

$$\frac{d}{d\tau_0} H^\mu(0) = 0, \tag{5.94}$$

and

$$\frac{d^2}{d\tau_0^2} H^\mu(0) = 2T^\mu|_{x_0}(v, w). \tag{5.95}$$

In words: Nonclosure of an infinitesimal geodesic parallelogram is proportional to its 'area' τ_0^2 and measured by the torsion 2-form. For any τ_0 (small enough) we define a linear mapping $G(\tau_0) \in GL(T_{x_0}\mathscr{U})$:

$$G(\tau_0): T_{x_0}\mathscr{U} \to T_{x_0}\mathscr{U}$$

$$r \mapsto G(\tau_0)r.$$

$G(\tau_0)\,r$ is the parallel transport of r along Q_1, Q_2, Q_3, Q_4 and backwards along H. Considered as a curve in the space of 4×4 matrices, $G(\tau_0)$ has the following properties:

$$\frac{d}{d\tau_0} G(0) = 0, \tag{5.96}$$

$$\frac{d^2}{d\tau_0^2} G(0) = 2R|_{x_0}(v, w) \in gl_4. \tag{5.97}$$

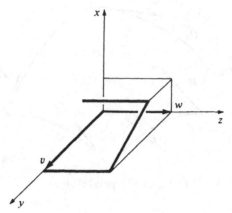

Fig. 5.2. $\mathbb{R}^3$ with torsion.

This is a slightly different formulation of the defining property of the field strength in the general case (chapter 4).

Curvature and torsion are local properties of the connection.

We have already seen a space with curvature and without torsion: a piece of the 2-sphere. We now discuss a space without curvature and with torsion. This example is due to E. Cartan and one might suspect a relation between torsion and spin. $\mathcal{U}$ is $\mathbb{R}^3$ and parallel transport of vectors is defined as follows: In the x- and y-directions they are transported by ordinary translations, while in the z-direction they are translated and at the same time rotated around the z-direction in the clockwise sense by an angle proportional to the z-displacement. This parallel transport is metric with respect to the Euclidean metric. A geodesic is either a straight line or a helix. The parallel transport has vanishing curvature because it is path independent. To see that it has torsion we construct a geodesic parallelogram with v in the y-direction and w in the z-direction (see fig. 5.2).

Problems

5.1 Derive the metric condition (5.23).

5.2 Show that the connection (5.51) is metric and torsion-free.

5.3 Consider a piece of the pseudosphere, that is an open subset of $\mathbb{R}^2$ endowed with the metric whose matrix with respect to a holonomic frame du, dv is

$$(g^{ij}) = \begin{pmatrix} e^v & 0 \\ 0 & 1 \end{pmatrix}$$

Calculate the Riemannian connection, curvature and curvature scalar.

5.4 Prove (5.78).

5.5 Give explicit expressions of the metric and the connection for Cartan's example, fig. 5.2. Calculate the curvature and the torsion.

6
The Lie derivative

This chapter deals with infinitesimal diffeomorphisms of an open subset $\mathcal{U}$ of $\mathbb{R}^n$. For the moment we do not need a metric on $\mathcal{U}$. In this case the geometric interpretation was to imagine $\mathcal{U}$ as a rubber sheet and a diffeomorphism as an active transformation deforming that sheet. We never embed $\mathcal{U}$ in some higher-dimensional $\mathbb{R}^N$. Consequently we do not see dimensions which are not 'in the plane' of $\mathcal{U}$ and we are sensitive only to deformations tangent to $\mathcal{U}$. It is no surprise then that an infinitesimal diffeomorphism is described by a vector field, a smooth collection of tangent vectors.

6.1 The flow of a vector field

Let Diff $(\mathcal{U})$ denote the group of diffeomorphisms from $\mathcal{U}$ onto itself. We define a one-parameter group of diffeomorphisms as a differentiable map

$$h: \mathbb{R} \times \mathcal{U} \to \mathcal{U}$$

$$(\tau, x) \mapsto h_\tau(x)$$

such that for fixed parameter τ, h_τ is a diffeomorphism of $\mathcal{U}$ satisfying

$$h_0(x) = x, \tag{6.1}$$

$$h_{\sigma+\tau}(x) = h_\sigma(h_\tau(x)). \tag{6.2}$$

In other words, the mapping $\tau \mapsto h_\tau$ is a group homomorphism from the additive group $\mathbb{R}$ into Diff $(\mathcal{U})$. In hydrodynamics, for instance, such one-parameter groups describe the time evolution of a fluid; e.g. think of a plastic bag floating in a river.

Many such time evolutions exist for only limited times. We describe this situation by local one-parameter groups of diffeomorphisms. They are defined only on open subsets $\mathcal{V} \subset \mathbb{R} \times \mathcal{U}$. We require that in every point $x \in \mathcal{U}$ an open interval of time is contained in $\mathcal{V}$,

$$\mathcal{V} \cap (\mathbb{R} \times \{x\}) = \text{interval} \subset \mathbb{R},$$

and we set our 'clock' such that

$$\{0\} \times \mathcal{U} \subset \mathcal{V}.$$

A local one-parameter group of diffeomorphisms is a differentiable map

$$h: \mathcal{V} \to \mathcal{U}$$

such that, as above,

$$h_0(x) = x \tag{6.3}$$

and

$$h_{\sigma + \tau}(x) = h_\sigma(h_\tau(x)) \tag{6.4}$$

whenever both sides are defined.

Each local one-parameter group h defines a vector field v: For fixed x, $\tau \mapsto h_\tau(x)$ is a curve in $\mathcal{U}$, e.g. the trajectory of a dirt particle in a river. The tangent vector of this curve at $\tau = 0$ is $v(x)$, the velocity. With respect to coordinates x^i:

$$v = \sum_{i=1}^{n} v^i \frac{\partial}{\partial x^i}, \tag{6.5}$$

$$v^i(x) := \frac{d}{d\tau} h_\tau^i(x)|_{\tau = 0}. \tag{6.6}$$

Conversely, each vector field v generates a local one-parameter group h, 'the flow of v', as solution to the differential equations

$$\frac{d}{d\tau} h_\tau^i(x) = v^i(h_\tau(x)). \tag{6.7}$$

For x fixed this is a system of ordinary, nonlinear, first-order differential equations, and together with the initial condition

$$h_0(x) = x \tag{6.8}$$

it always has a unique solution $h_\tau(x)$ for small enough τ. The uniqueness then implies (6.4) because both sides are solutions. In general we have to restrict ourselves to small times τ such that the flow h_τ does not leave $\mathcal{U}$.

Here is an example where h is defined for all times: Let $\mathcal{U} = \mathbb{R}^n$ and v be a linear vector field

$$v(x) = Ax, \quad A \in gl_n, \tag{6.9}$$

then

$$h_\tau(x) = e^{\tau A} x. \tag{6.10}$$

6.2 Lie derivative of forms

Diff($\mathcal{U}$) is an (infinite-dimensional) continuous group. We have already encountered a linear representation ρ of Diff($\mathcal{U}$). It was defined on the (infinite-dimensional) vector space $\Lambda\mathcal{U}$ of differential forms and given by the pullback (see equation (2.40)):

$$\rho: \mathrm{Diff}(\mathcal{U}) \to GL(\Lambda\mathcal{U})$$

$$F \mapsto \rho_F,$$

$$\rho_F := (F^{-1})^*. \tag{6.11}$$

We now define the corresponding representation of a vector field v, the Lie derivative. Let h be the flow of v. The Lie derivative $L_v\varphi$ of a p-form φ in direction v is again a p-form:

$$L_v: \Lambda^p\mathcal{U} \to \Lambda^p\mathcal{U}$$

$$\varphi \mapsto L_v\varphi,$$

$$L_v\varphi := \frac{d}{d\tau} h_\tau^* \varphi|_{\tau=0}. \tag{6.12}$$

It has the following immediate properties:

(a) L_v is a linear mapping;
(b) Leibniz rule: $L_v(\varphi \wedge \psi) = (L_v\varphi) \wedge \psi + \varphi \wedge L_v\psi;$ (6.13)
(c) $dL_v = L_v d$ (for $dF^* = F^*d$);   (6.14)
(d) L_v is linear in v.

For example, if $f \in \Lambda^0\mathcal{U}$ is a function,

$$(L_v f)(x) = \lim_{\tau \to 0} \frac{1}{\tau} [f(h_\tau(x)) - f(x)]$$

$$= \lim_{\tau \to 0} \frac{1}{\tau} [f(x + \tau v(x)) - f(x)] = \sum_i v^i \frac{\partial}{\partial x^i} f$$

$$= df|_x(v) = i_v df(x). \tag{6.15}$$

In words: The Lie derivative of a function is the directional derivative. Remembering the notation of v in coordinates,

$$v = \sum_i v^i \frac{\partial}{\partial x^i},$$

the directional derivative is sometimes written

$$v(f):= \sum_i v^i \frac{\partial}{\partial x^i} f \qquad (6.16)$$

and read: 'v applied to f'.

6.3 The Lie bracket

Let vect$(\mathcal{U})$ denote the infinite-dimensional vector space of all vector fields on $\mathcal{U}$. As infinitesimal elements of the continuous group Diff$(\mathcal{U})$ they form a Lie algebra. Indeed we can define the bracket of two vector fields v and w. If in coordinates

$$v = \sum_i v^i \frac{\partial}{\partial x^i}, \quad w = \sum_j w^j \frac{\partial}{\partial x^j}, \qquad (6.17)$$

we define the components of $[v, w]$ by

$$[v, w]:= \sum_{i,j} \left(v^i \frac{\partial}{\partial x^i} w^j - w^i \frac{\partial}{\partial x^i} v^j \right) \frac{\partial}{\partial x^j}. \qquad (6.18)$$

This definition is independent of the choice of coordinates and has the following properties:

(a) it is bilinear;
(b) it is antisymmetric:

$$[v, w] = -[w, v]; \qquad (6.19)$$

(c) the Jacobi identity:

$$[v, [u, w]] = [[v, u], w] + [u, [v, w]]. \qquad (6.20)$$

Remember that a vector space (finite-dimensional or not) together with a product $[\ , \ ]$ satisfying (a), (b), and (c) is called a Lie algebra. Therefore vect$(\mathcal{U})$ is an infinite-dimensional Lie algebra, the Lie algebra of Diff$(\mathcal{U})$. The Lie derivative defines a linear representation of this Lie algebra on the vector space $\Lambda^0 \mathcal{U}$:

$$[L_v, L_w]f = L_{[v,w]}f. \qquad (6.21)$$

But we have more than just a representation. The Lie derivative obeys the Leibniz rule:

$$L_v(fg) = (L_v f)g + f L_v g \qquad (6.22)$$

for all $v \in$ vect$(\mathcal{U})$ and $f, g \in \Lambda^0 \mathcal{U}$. Note that $\Lambda^0 \mathcal{U}$ together with pointwise multiplication forms an algebra denoted by $C^\infty(\mathcal{U})$. One defines: A linear mapping D of an algebra $\mathfrak{a}$ into itself is called derivation if it satisfies

$$D(fg) = (Df)g + f\,Dg \qquad (6.23)$$

for all $f, g \in \mathfrak{a}$. The set of derivations on $\mathfrak{a}$, der $\mathfrak{a}$, is also a Lie algebra with bracket

$$[D_1, D_2]f := D_1 D_2 f - D_2 D_1 f. \qquad (6.24)$$

Now for each derivation D on $C^\infty(\mathcal{U})$ there is a unique vector field v such that $D = L_v$. This statement, which is not easy to prove, motivates the peculiar notation

$$v = \sum_i v^i \frac{\partial}{\partial x^i} \qquad (6.25)$$

and allows a more abstract definition of vector fields as derivations instead of the more pedestrian definition (2.9) using coordinates.

To summarize, the map

$$\text{vect}(\mathcal{U}) \to \text{der}\, C^\infty(\mathcal{U})$$

$$v \mapsto L_v$$

is a Lie algebra isomorphism.

6.4 Derivations of forms

This section is another variation on a theme by Leibniz generalizing the above from 0-forms to any p-form.

We call a linear map

$$D: \Lambda^p \mathcal{U} \to \Lambda^{p+d}\mathcal{U}$$

a graded derivation on $\Lambda \mathcal{U}$ of degree d, $d \in \mathbb{Z}$, if it satisfies:

$$D(\varphi \wedge \psi) = (D\varphi) \wedge \psi + (-1)^{d\deg\varphi}\varphi \wedge D\psi. \qquad (6.26)$$

We have already seen three examples: the inner derivatives of degree -1, the Lie derivatives of degree 0, and the exterior derivative of degree 1.

The set of all graded derivations of $\Lambda \mathcal{U}$ is an infinite-dimensional graded Lie algebra with bracket:

$$[D_1, D_2] := D_1 D_2 - (-1)^{d_1 d_2} D_2 D_1. \qquad (6.27)$$

A graded (or super) Lie algebra is a collection of vector spaces indexed by the integers, the degree, with a bracket [,] satisfying the following properties:

(a) it is bilinear;
(b) it is graded commutative:

$$[D_1, D_2] = -(-1)^{d_1 d_2}[D_2, D_1];$$ (6.28)

(c) the graded Jacobi identity

$$[D_1, [D_2, D_3]] = [[D_1, D_2], D_3] + (-1)^{d_1 d_2}[D_2, [D_1, D_3]].$$ (6.29)

The inner, Lie, and exterior derivatives together form an (infinite-dimensional) graded Lie subalgebra. Indeed we have the following commutation relations, which will prove helpful in later applications:

$$[L_v, L_w] = L_v L_w - L_w L_v = L_{[v,w]},$$ (6.30)

$$[i_v, i_w] = i_v i_w + i_w i_v = 0,$$ (6.31)

$$[d, d] = d^2 + d^2 = 0,$$ (6.32)

$$[L_v, d] = L_v d - d L_v = 0,$$ (6.33)

$$[i_v, d] = i_v d + d i_v = L_v,$$ (6.34)

$$[L_v, i_w] = L_v i_w - i_w L_v = i_{[v,w]}.$$ (6.35)

Equation (6.34) is due to Henri Cartan. On 0-forms it reduces to equation (6.15).

As a side remark concerning the case where $\mathcal{U}$ is equipped with an orientation and a metric, we mention that the coderivative δ is not a derivation.

6.5 An application in general relativity

Any Lagrangian theory formulated in terms of differential forms is invariant under diffeomorphisms. In particular for gauge theories, like Yang–Mills theories and general relativity, the invariance group is the semidirect product of the diffeomorphisms and the gauge group. Einstein, by fixing the gauge, reduced the gauge group $^\ast GL_4^+$ to a substructure of $^\ast GL_4^+$, the coordinate transformations, In some respects these appear similar to diffeomorphisms, the other factor in the semidirect product. (Their distinction will become clearer in the context of manifolds.) As in the literature, diffeomorphisms and coordinate transformations are frequently mixed up and in order to make clear that diffeomorphisms are of secondary importance in the construction of general relativity, we have postponed this discussion until now, even though logically it belongs to chapter 2. Remember that we have encountered (infinitesimal) diffeomorphisms only once in the last chapter, when we wanted to prove the covariant conservation of energy–momentum of matter. We now give the details.

We start from the remark that any action $\int_{\mathcal{U}} \mathcal{L}$, the integral of a Lagrangian over all spacetime, is invariant under infinitesimal diffeomorphisms. Of course we suppose that $\mathcal{L}$ is a 4-form with appropriate asymptotic behaviour such that its integral converges.

Indeed,

$$
\int_{\mathcal{U}} L_v \mathcal{L} = \int_{\mathcal{U}} (i_v d + d i_v) \mathcal{L} = \int_{\mathcal{U}} d i_v \mathcal{L}
$$

$$
= \int_{\partial \mathcal{U}} i_v \mathcal{L} = 0 \quad \text{for all } v \in \text{vect}(\mathcal{U}). \tag{6.36}
$$

The first equality is Cartan's formula (6.34), the second holds because $d\mathcal{L} = 0$ as a 5-form in a four-dimensional space, the third is Stokes' theorem, and the fourth equality follows from the asymptotic behaviour of $\mathcal{L}$.

Varying the frame in the matter Lagrangian with zero torsion and the matter fields on shell, we get, using the definition of τ, equation (5.30),

$$
0 = -\int (L_v e^c) \wedge \tau_c
$$

$$
= -\int (d i_v e^c) \wedge \tau_c - \int (i_v d e^c) \wedge \tau_c
$$

$$
= \int (i_v e^c) d\tau_c + \int i_v (\omega^c{}_a \wedge e^a) \wedge \tau_c
$$

$$
= \int (i_v e^c) d\tau_c + \int (i_v \omega^c{}_a) e^a \wedge \tau_c - \int \omega^c{}_a \wedge (i_v e^a) \wedge \tau_c
$$

$$
= \int (i_v e^c)(d\tau_c + \omega_c{}^a \wedge \tau_a) + \int (i_v \omega_{ca}) e^a \wedge \tfrac{1}{6} \tau^{cd} \varepsilon_{drst} e^r \wedge e^s \wedge e^t
$$

$$
= \int (i_v e^c)(D\tau)_c. \tag{6.37}
$$

The last term before the last equal sign vanishes due to the antisymmetry of ω_{ca} and symmetry of τ^{ca}. The $i_v e^c$ are the components of v with respect to the orthonormal frame e^a and therefore arbitrary implying the covariant conservation of τ.

6.6 Lie derivative of vector fields

We start from the linear representation of $\mathrm{Diff}(\mathcal{U})$ on the vector fields,

$$\mathrm{Diff}(\mathcal{U}) \to GL(\mathrm{vect}(\mathcal{U}))$$

$$F \mapsto TF,$$

and write down its infinitesimal version.

Let v be a vector field, h its flow. We define the Lie derivative of a vector field w in the direction v:

$$L_v : \mathrm{vect}(\mathcal{U}) \to \mathrm{vect}(\mathcal{U})$$

$$w \mapsto L_v w,$$

$$(L_v w)(x) := \frac{\mathrm{d}}{\mathrm{d}\tau} T_{h_\tau(x)}(h_\tau^{-1}) w(h_\tau(x))|_{\tau=0}. \tag{6.38}$$

An easy calculation shows that

$$L_v w = [v, w]. \tag{6.39}$$

L_v, the infinitesimal form of the above representation of $\mathrm{Diff}(\mathcal{U})$, is the adjoint representation of $\mathrm{vect}(\mathcal{U})$ on itself. Note the peculiar way we have written the Jacobi identity (6.20). It can now be read: v acts on the product $[u, w]$ like a derivation.

6.7 Killing vectors

We now endow our open subset $\mathcal{U}$ with a metric g and consider the isometries of g. They form a subgroup of $\mathrm{Diff}(\mathcal{U})$ and we are interested in its Lie algebra.

Let h be a one-parameter group of isometries. Its vector field

$$v^i(x) = \frac{\mathrm{d}}{\mathrm{d}\tau} h_\tau^i(x)|_{\tau=0} \tag{6.40}$$

is called a Killing vector (field). It follows that v is a Killing vector if and only if

$$L_v g(u, w) = g(L_v u, w) + g(u, L_v w) \tag{6.41}$$

for all vector fields u and w.

Expressed with respect to an orthonormal frame e_i the Lie derivative is written as

$$L_v e_i(x) = \sum_k a^k{}_i(x) e_k(x) \tag{6.42}$$

and (6.41) becomes

$$L_v \eta_{ij} = L_v g(e_i, e_j)$$

$$= g\left(\sum_k a^k{}_i e_k, e_j\right) + g\left(e_i, \sum_k a^k{}_j e_k\right)$$

$$= a_{ji} + a_{ij} = 0. \tag{6.43}$$

Without proof we cite the theorem: The Killing vectors are a finite-dimensional subalgebra of vect$(\mathcal{U})$ with dimension smaller than or equal to $\frac{1}{2}n(n+1)$.

In general, a metric has no Killing vector different from zero; a space with a metric is called maximally symmetric if the dimension of its Lie algebra of Killing vectors is equal to $\frac{1}{2}n(n+1)$.

As an example take $\mathcal{U} = \mathbb{R}^2$ with Euclidean metric. The vector fields $\partial/\partial x$, $\partial/\partial y$ (infinitesimal translations) and $x(\partial/\partial y) - y(\partial/\partial x)$ (infinitesimal rotation) are three independent Killing vectors; they generate the Euclidean group.

Note that the isometries in flat space physics play a privileged role, e.g. the Poincaré invariance of Maxwell's equations in flat space (cf. the discussion around equation (4.11)). In general these symmetries disappear when we go to a curved space with fixed metric and *a fortiori* in general relativity where the metric becomes a dynamical variable.

Problems

6.1 Verify the Jacobi identity (6.20) for the Lie bracket.

6.2 Prove Cartan's identity (6.34).

6.3 Consider a piece of the 2-sphere with its standard metric (5.55). Show that the following three vector fields

$$\frac{\partial}{\partial \varphi},$$

$$\cos\varphi \frac{\cos\vartheta}{\sin\vartheta}\frac{\partial}{\partial\varphi} + \sin\varphi\frac{\partial}{\partial\vartheta},$$

$$-\sin\varphi\,\frac{\cos\vartheta}{\sin\vartheta}\,\frac{\partial}{\partial\varphi}+\cos\varphi\,\frac{\partial}{\partial\vartheta}$$

are Killing vectors. What Lie algebra do they generate?

7

Manifolds

In this chapter we want to generalize the concepts introduced so far to a class of spaces, the differentiable manifolds, which in general cannot be considered as open subsets of $\mathbb{R}^n$. An example of such a space is the 2-sphere, e.g. the surface of the Earth: The globe as a whole cannot be mapped homeomorphically onto an open subset of the Euclidean plane. Instead, one has to be content with local mappings collected to make an atlas. If it is indicated how points in the overlap of different maps are to be identified, e.g. by means of parallels and meridians, one gets nevertheless a complete description of the surface of the Earth. Note that there is a lot of freedom connected with these local mappings: One can choose among a variety of map projections.

In chapter 2 we already encountered surfaces which could not be described by a single parameter representation. Since we used them only as domains of integration, the details of the gluing procedure were irrelevant. In this chapter, however, we shall be more careful about the transition from one parameter representation to another.

7.1 Differentiable manifolds

An n-dimensional manifold is a topological space M which locally looks like an open subset of $\mathbb{R}^n$. To make the 'looks like' more precise, we introduce the following notion: A chart (also called a local coordinate system) for M is a pair $(\mathcal{U}, \alpha)$, where $\mathcal{U}$ is an open subset of M and where α is a homeomorphism of $\mathcal{U}$ onto an open subset of $\mathbb{R}^n$,

$$\alpha: \mathcal{U} \to \alpha(\mathcal{U}); \tag{7.1}$$

i.e. α is bijective with α and α^{-1} continuous. The homeomorphism α identifies $\mathcal{U}$ with $\alpha(\mathcal{U})$ assigning to each $x \in \mathcal{U}$ an n-tuple of real numbers $x^1 = \alpha^1(x), x^2 = \alpha^2(x), \ldots, x^n = \alpha^n(x)$, the coordinates of x with respect to the chart $(\mathcal{U}, \alpha)$. The open set $\mathcal{U} \subset M$ is called a coordinate neighbourhood.

Note that the x^i actually have two meanings. They may be considered as coordinates of $x \in M$ and as the Cartesian coordinates of $\alpha(x) \in \mathbb{R}^n$. If we do not want to mention the homeomorphism α explicitly, we shall write $x^i(x)$ instead of $x^i = \alpha^i(x)$.

Now we are ready to give the definition of a topological manifold. It is a Hausdorff space M with a countable basis for its topology such that every point of M lies in a coordinate neighbourhood $\mathscr{U}$; i.e. M is covered by the $\mathscr{U}$s. Consequently, M can be described locally by the corresponding coordinates. Note that our assumption of a countable basis for the topology of M implies that M is paracompact. Do not worry if you are not familiar with these topological concepts. They are only mentioned in order to formulate precisely the hypotheses under which the following constructions work and will not appear again.

Consider now two charts $(\mathscr{U}_1, \alpha_1)$, $(\mathscr{U}_2, \alpha_2)$ with $\mathscr{U}_1 \cap \mathscr{U}_2 \neq \varnothing$. So each $x \in \mathscr{U}_1 \cap \mathscr{U}_2$ possesses two sets of coordinates:

$$x^i = \alpha_1^i(x), \quad y^i = \alpha_2^i(x). \tag{7.2}$$

The relation between the y-coordinates and the x-coordinates is given by the map $\alpha_2 \circ \alpha_1^{-1}$:

$$y^i = \alpha_2^i(x) = \alpha_2^i(\alpha_1^{-1}(x^1, x^2, \ldots, x^n)); \tag{7.3}$$

i.e.

$$(y^1, y^2, \ldots, y^n) = \alpha_2 \circ \alpha_1^{-1}(x^1, x^2, \ldots, x^n). \tag{7.4}$$

This map is a homeomorphism of open subsets of $\mathbb{R}^n$ (see fig. 7.1):

$$\alpha_2 \circ \alpha_1^{-1} : \alpha_1(\mathscr{U}_1 \cap \mathscr{U}_2) \to \alpha_2(\mathscr{U}_1 \cap \mathscr{U}_2). \tag{7.5}$$

It describes how the different charts are to be glued together. As a shorthand for the change of coordinates (also called coordinate transformation) (7.3) we shall often write

$$y^i = y^i(x^1, x^2, \ldots, x^n).$$

A collection of charts $\{(\mathscr{U}_r, \alpha_r)\}$ such that the corresponding coordinate neighbourhoods cover M, i.e. $\cup_r \mathscr{U}_r = M$, is called an atlas for the topological manifold M. The changes of coordinates $\alpha_s \circ \alpha_r^{-1}$ with nonempty domains of definition are homeomorphisms of open subsets of $\mathbb{R}^n$. If the charts satisfy the additional compatibility condition that the maps $\alpha_s \circ \alpha_r^{-1}$ are not only homeomorphisms but also diffeomorphisms, the atlas is called a C^∞-atlas.

By definition, a differentiable (C^∞) structure on a topological manifold M is a maximal C^∞-atlas, i.e. a C^∞-atlas containing all charts for M which are compatible in the above sense. A topological manifold equipped with

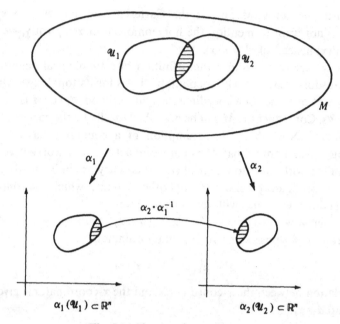

Fig. 7.1. Change of coordinates.

a differentiable structure is called a differentiable (C^∞) manifold.

In an analogous manner one defines C^k-manifolds for $1 \leqslant k < \infty$ and analytic manifolds. Replacing in the above definitions $\mathbb{R}^n$ by $\mathbb{C}^n$ and C^∞ by holomorphic, one arrives at the notion of a complex manifold. Since we shall consider only C^∞-manifolds, we shall speak of them simply as manifolds, and 'differentiable' or 'smooth' will always mean infinitely often differentiable, i.e. C^∞.

The most trivial example of a differentiable manifold is an open subset $\mathcal{U}$ of $\mathbb{R}^n$. An atlas consisting of just one chart is $\{(\mathcal{U}, \mathrm{id}_{\mathcal{U}})\}$.

Another example is provided by the n-dimensional sphere

$$S^n = \left\{ (x^1, x^2, \ldots, x^{n+1}) \in \mathbb{R}^{n+1} \,\middle|\, \sum_{i=1}^{n+1} (x^i)^2 = 1 \right\}. \tag{7.6}$$

Put

$$\mathcal{U}_1 := \{ (x^1, x^2, \ldots, x^{n+1}) \in S^n \,|\, x^{n+1} > -1 \}$$

$$= S^n - \{\text{south pole}\}, \tag{7.7}$$

$$\mathcal{U}_2 := \{ (x^1, x^2, \ldots, x^{n+1}) \in S^n \,|\, x^{n+1} < 1 \}$$

$$= S^n - \{\text{north pole}\} \tag{7.8}$$

and

$$\alpha_1(x^1,x^2,\ldots,x^{n+1}):=\frac{1}{1+x^{n+1}}(x^1,x^2,\ldots,x^n),\quad x^{n+1}\neq -1, \quad (7.9)$$

$$\alpha_2(x^1,x^2,\ldots,x^{n+1}):=\frac{1}{1-x^{n+1}}(x^1,x^2,\ldots,x^n),\quad x^{n+1}\neq 1. \quad (7.10)$$

The coordinate mappings α_1, α_2 correspond to stereographic projection from the south pole and the north pole, respectively. Since the change of coordinates

$$\alpha_1\circ\alpha_2^{-1}(x^1,x^2,\ldots,x^n)=\left(\sum_{i=1}^{n}(x^i)^2\right)^{-1}(x^1,x^2,\ldots,x^n) \quad (7.11)$$

is infinitely often differentiable on $\alpha_2(\mathcal{U}_1\cap\mathcal{U}_2)=\mathbb{R}^n-\{0\}$, the set of two charts $\{(\mathcal{U}_1,\alpha_1),(\mathcal{U}_2,\alpha_2)\}$ is a C^∞-atlas. It determines a differentiable structure for S^n.

Any open subset N of a manifold M possesses a natural differentiable structure: Take as charts for N the pairs $(\mathcal{U}_r\cap N,\alpha_r|_{\mathcal{U}_r\cap N})$, where $\{(\mathcal{U}_r,\alpha_r)\}$ is a C^∞-atlas for M.

7.2 Differentiable mappings

On a differentiable manifold M we can define differentiable functions and related notions.

A function

$$f:M\to\mathbb{R} \quad (7.12)$$

is called differentiable at $x\in M$, if among the charts of the differentiable structure there is one chart $(\mathcal{U},\alpha)$ with $x\in\mathcal{U}$ such that

$$f\circ\alpha^{-1}:\alpha(\mathcal{U})\to\mathbb{R} \quad (7.13)$$

is differentiable (in the sense of ordinary calculus) at $\alpha(x)$. This means the coordinate representation $f(x^1,x^2,\ldots,x^n)$ has to be a differentiable function.

Of course, the definition should be independent of the choice of the chart $(\mathcal{U},\alpha)$. This is easily checked. Let $(\mathcal{V},\beta)$ be another chart from our differentiable structure with $x\in\mathcal{V}$. Then we have

$$f\circ\beta^{-1}=(f\circ\alpha^{-1})\circ(\alpha\circ\beta^{-1}). \quad (7.14)$$

Since $\alpha\circ\beta^{-1}$ is a diffeomorphism, $f\circ\alpha^{-1}$ is differentiable at $\alpha(x)$ if and only if $f\circ\beta^{-1}$ is differentiable at $\beta(x)$.

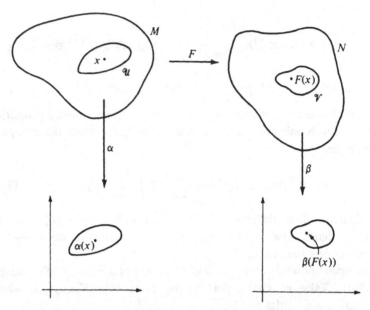

Fig. 7.2. Coordinate representation of $F: M \to N$.

The algebra of functions $f: M \to \mathbb{R}$ which are differentiable at each $x \in M$ is denoted by $C^\infty(M)$.

More generally, we consider maps

$$F: M \to N \tag{7.15}$$

where M and N are manifolds with dim $M = n$, dim $N = p$. Let $(\mathcal{U}, \alpha)$ be a chart for M, $x \in \mathcal{U}$, and $(\mathcal{V}, \beta)$ a chart for N, $F(x) \in \mathcal{V}$. The map F is called differentiable at x if $\beta \circ F \circ \alpha^{-1}$ is differentiable at $\alpha(x)$ (see fig. 7.2). The mapping $\beta \circ F \circ \alpha^{-1}$, i.e. the functions $F^j(x^1, x^2, \ldots, x^n)$, $j = 1, 2, \ldots, p$, represent F in local coordinates. So F is differentiable at x if the coordinates of $F(x)$ are differentiable functions of the coordinates of x. The independence of this definition of the chosen coordinates is shown as for real-valued functions. If F is differentiable at each $x \in M$, it is simply called differentiable.

From these considerations we can deduce a general recipe to translate local concepts from $\mathbb{R}^n$ to manifolds: Since coordinate systems identify a manifold locally with an open subset of $\mathbb{R}^n$, the use of coordinate representations of the objects involved reduces everything to the familiar case of $\mathbb{R}^n$. Of course, this procedure makes sense only for notions which are invariant under coordinate transformations (diffeomorphisms of open subsets of $\mathbb{R}^n$). Otherwise our definitions would depend on the coordinates.

Take as an example a differentiable map F as in (7.15). Then we can define the rank of F at $x \in M$ to be the rank of the coordinate representation $\beta \circ F \circ \alpha^{-1}$ at $\alpha(x)$, i.e. the rank of the matrix

$$\left(\frac{\partial F^j}{\partial x^i} \right)_{\substack{j=1,2,\ldots,p \\ i=1,2,\ldots,n}}.$$

Global objects may be constructed by gluing together local coordinate representations. One defines them on coordinate neighbourhoods $\mathcal{U}_r$, trying to arrange the definitions such that they coincide on the intersections $\mathcal{U}_r \cap \mathcal{U}_s$. Of course, this will not always be possible. If not, the corresponding object does not exist on the manifold considered, such as for example an orientation on the Möbius strip. (The concept of orientation will be explained below.)

Having defined differentiable maps of manifolds, it is obvious how to define diffeomorphisms. The map (7.15) is a diffeomorphism if it is bijective and F as well as F^{-1} are differentiable. Then $\dim M = \dim N$, and we call M and N diffeomorphic. Note that a diffeomorphism of two manifolds is globally defined: It maps the whole manifold M onto the whole manifold N. The diffeomorphisms of M onto itself obviously form a group denoted by $\mathrm{Diff}(M)$.

In particular, let $(\mathcal{U}, \alpha)$ be a chart for M. Then α is a diffeomorphism from $\mathcal{U}$ to $\alpha(\mathcal{U})$ if $\alpha(\mathcal{U})$ is given the standard differentiable structure of open subsets of $\mathbb{R}^n$ as explained above.

Consider now the following two manifolds. Let M be $\mathbb{R}$ with the standard topology and the differentiable structure defined by the atlas $\{(\mathbb{R}, \alpha)\}$, where $\alpha(x) = x$ for all $x \in \mathbb{R}$. Take $\mathbb{R}$ again for N with the standard topology, but equipped with the differentiable structure given by the atlas $\{(\mathbb{R}, \beta)\}$, where $\beta(x) = x^{1/3}$ for all $x \in \mathbb{R}$. These two differentiable structures on $\mathbb{R}$ are not equivalent, i.e. the charts $(\mathbb{R}, \alpha)$ and $(\mathbb{R}, \beta)$ are not compatible, because $\beta \circ \alpha^{-1}(x) = x^{1/3}$ so that $\beta \circ \alpha^{-1}$ is not everywhere differentiable. Consequently, one can have different differentiable structures on one topological manifold. On the other hand, there are topological manifolds on which no differentiable structure can be defined.

The fact that M and N are distinct as differentiable manifolds has the consequence that $C^\infty(M) \neq C^\infty(N)$. Take, for example, the function

$$f : \mathbb{R} \to \mathbb{R}$$
$$x \mapsto x^{1/3}.$$

(7.16)

It is easy to see that $f \notin C^\infty(M)$, but $f \in C^\infty(N)$. Nevertheless, M and N are diffeomorphic: The map

$$F: M \to N$$
$$x \mapsto x^3$$

(7.17)

is a diffeomorphism as one immediately confirms.

Are there differentiable structures on $\mathbb{R}$, or more generally on $\mathbb{R}^n$, such that the corresponding manifolds are not diffeomorphic? The following theorem answers this question. On $\mathbb{R}^n$ with the standard topology, $n \in \mathbb{N}$, $n \neq 4$, all differentiable structures lead to diffeomorphic manifolds. However, on $\mathbb{R}^4$ there are different differentiable structures defining non-diffeomorphic manifolds (Freed & Uhlenbeck 1984).

7.3 The Cartesian product of manifolds

On the Cartesian product of two manifolds a differentiable structure can be constructed in the following way. Let M be an n-dimensional manifold with atlas $\{(\mathcal{U}_r, \alpha_r)\}$ and N a p-dimensional manifold with atlas $\{(\mathcal{V}_s, \beta_s)\}$. Define coordinate maps $\alpha_r \times \beta_s$ for $M \times N$:

$$\alpha_r \times \beta_s : \mathcal{U}_r \times \mathcal{V}_s \to \mathbb{R}^n \times \mathbb{R}^p = \mathbb{R}^{n+p}$$
$$(x, y) \mapsto (\alpha_r(x), \beta_s(y)).$$

(7.18)

The atlas $\{(\mathcal{U}_r \times \mathcal{V}_s, \alpha_r \times \beta_s)\}$ determines a differentiable structure for the $(n + p)$-dimensional manifold $M \times N$. In the following, the Cartesian product of differentiable manifolds will always be assumed to be equipped with this differentiable structure. So, for example, the n-dimensional torus $T^n = S^1 \times S^1 \times \cdots \times S^1$ (n factors S^1) becomes a differentiable manifold.

7.4 Submanifolds

A subset N of an n-dimensional manifold M is called a q-dimensional submanifold of M, if for each point $x_0 \in N$ there is a chart $(\mathcal{U}, \alpha)$ for M with $x_0 \in \mathcal{U}$ such that

$$\alpha(x) = (x^1, x^2, \ldots, x^q, a^1, a^2, \ldots, a^{n-q})$$

(7.19)

for all $x \in \mathcal{U} \cap N$ with a^i fixed (see fig. 7.3). For such a chart define

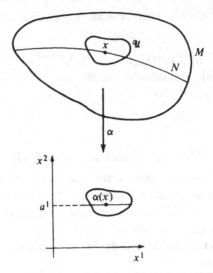

Fig. 7.3. The definition of a submanifold.

$\bar{\mathcal{U}} := \mathcal{U} \cap N$ and $\bar{\alpha}: \bar{\mathcal{U}} \to \mathbb{R}^q$ by

$$\bar{\alpha}(x) = (x^1, x^2, \ldots, x^q). \tag{7.20}$$

Then $(\bar{\mathcal{U}}, \bar{\alpha})$ is a chart for N, and the collection of all these charts forms an atlas determining a differentiable structure for N. One should be aware of the fact that there are several different definitions of 'submanifold' in current use. Because of its simplicity we chose this one, which is relatively restrictive.

A trivial example of a submanifold (with $q = n$) is provided by any open subset of M. Furthermore, finite sets of points of M may be considered as zero-dimensional submanifolds.

In many cases one encounters subsets of a manifold which are defined by systems of equations. Think, for example, of constraints in classical mechanics determining the configuration space of a mechanical system. The following theorem gives a convenient criterion to decide when such subsets are actually submanifolds.

THEOREM

Let $N := \{x \in M \mid f^A(x) = 0, A = 1, 2, \ldots, p\}$, where $f^A \in C^\infty(M)$ such that the map

$$M \to \mathbb{R}^p \tag{7.21}$$

$$x \mapsto (f^1(x), f^2(x), \ldots, f^p(x))$$

has rank p for each $x \in N$. Then N is an $(n-p)$-dimensional submanifold of M.

For instance, $S^{n-1} = \{(x^1, x^2, \ldots, x^n) \in \mathbb{R}^n \mid \sum_{i=1}^{n}(x^i)^2 - 1 = 0\}$ is an $(n-1)$-dimensional submanifold of $\mathbb{R}^n$ with the same differentiable structure as introduced above.

7.5 Tangent space and vector fields

In chapter 6 we have seen that vector fields on open subsets of $\mathbb{R}^n$ can be identified with derivations (directional derivatives) of differentiable functions. This identification serves as the starting point for the definition of tangent vectors on a manifold M.

Let $(\mathcal{U}, \alpha)$ be a chart for M and denote the corresponding coordinates by $x^1, x^2, \ldots, x^n$. Let

$$f : \mathcal{U} \to \mathbb{R} \tag{7.22}$$

be differentiable at $x \in \mathcal{U}$. Remember that this means that

$$f \circ \alpha^{-1} : \alpha(\mathcal{U}) \to \mathbb{R} \tag{7.23}$$

is differentiable at $\alpha(x)$. Now define operators $\partial/\partial x^i|_x$, which act on differentiable functions, by

$$\left.\frac{\partial}{\partial x^i}\right|_x f := \left.\frac{\partial}{\partial x^i}(f \circ \alpha^{-1})\right|_{\alpha(x)} = \frac{\partial}{\partial x^i} f(x^1, x^2, \ldots, x^n). \tag{7.24}$$

They satisfy the Leibniz rule:

$$\left.\frac{\partial}{\partial x^i}\right|_x (fg) = \left(\left.\frac{\partial}{\partial x^i}\right|_x f\right) g(x) + f(x) \left.\frac{\partial}{\partial x^i}\right|_x g, \tag{7.25}$$

where f and g are differentiable at x. The operator $\partial/\partial x^i|_x$, also denoted by $\partial/\partial x^i(x)$, is the derivative along the x^i-coordinate line at x.

The real vector space spanned by

$$\frac{\partial}{\partial x^1}(x), \frac{\partial}{\partial x^2}(x), \ldots, \frac{\partial}{\partial x^n}(x)$$

is called the tangent space of M at x and denoted by $T_x M$. So any tangent vector $v \in T_x M$ may be written as

$$v = \sum_{i=1}^{n} v^i \frac{\partial}{\partial x^i}(x) \tag{7.26}$$

with $v^i \in \mathbb{R}$. It is an operator which acts on functions differentiable at x and obeys the Leibniz rule. The special tangent vectors

$$\frac{\partial}{\partial x^1}(x), \frac{\partial}{\partial x^2}(x), \ldots, \frac{\partial}{\partial x^n}(x)$$

form the coordinate basis or holonomic basis of $T_x M$ associated with the coordinates x^i.

With respect to another chart, whose coordinates are $y^1, y^2, \ldots, y^n$, we have

$$v = \sum_{j=1}^{n} w^j \frac{\partial}{\partial y^j}(x). \tag{7.27}$$

By means of the chain rule we find (cf. (2.14))

$$\frac{\partial}{\partial x^i}(x) = \sum_{j=1}^{n} \frac{\partial y^j}{\partial x^i} \frac{\partial}{\partial y^j}(x), \tag{7.28}$$

where $y^j(x^1, x^2, \ldots, x^n)$ denotes the change of coordinates, and consequently we get

$$w^j = \sum_{i=1}^{n} \frac{\partial y^j}{\partial x^i} v^i, \tag{7.29}$$

the well-known transformation rule for the components of a tangent vector with respect to a holonomic basis.

By analogy with (2.5) we define the tangent bundle of the manifold M as

$$TM := \bigcup_{x \in M} T_x M. \tag{7.30}$$

Now we let the point x in (7.26) vary over $\mathcal{U}$ and get a (smooth) vector field v on $\mathcal{U}$, i.e. a linear combination

$$v = \sum_{i=1}^{n} v^i \frac{\partial}{\partial x^i}, \tag{7.31}$$

where each v^i is a smooth function on $\mathcal{U}$. So v is a map

$$v: \mathcal{U} \to \bigcup_{x \in \mathcal{U}} T_x M \tag{7.32}$$

with

$$v(x) = \sum_{i=1}^{n} v^i(x) \frac{\partial}{\partial x^i}(x) \in T_x M. \tag{7.33}$$

A vector field on the whole manifold M is given by vector fields v, on $\mathcal{U}$, such that $v_r = v_s$ on $\mathcal{U}_r \cap \mathcal{U}_s$, where $\{(\mathcal{U}_r, \alpha_r)\}$ is an atlas for M. More explicitly, if $x_r^1, x_r^2, \ldots, x_r^n$ are the coordinates on $\mathcal{U}_r$, and

$$v_r = \sum_{i=1}^n v_r^i \frac{\partial}{\partial x_r^i}, \tag{7.34}$$

we must have, according to (7.29),

$$v_s^j = \sum_{i=1}^n \frac{\partial x_s^j}{\partial x_r^i} v_r^i. \tag{7.35}$$

Similarly we can describe vector fields defined on subsets of M. The set of all smooth vector fields on M is denoted by vect (M).

If $v \in$ vect (M) and $f \in C^\infty(M)$, we define a function $v(f) \in C^\infty(M)$ by

$$v(f)(x) = v(x)(f), \tag{7.36}$$

remembering that $v(x) \in T_x M$ is a directional derivative operating on elements of $C^\infty(M)$. Therefore, v can be interpreted as a derivation on $C^\infty(M)$ (cf. section 6.3). Conversely, any derivation on $C^\infty(M)$ stems from a vector field in the way described above.

The Lie bracket $[v, w]$ of two vector fields v and w is given by the following derivation (cf. (6.21)):

$$[v, w](f) = v(w(f)) - w(v(f)), \quad f \in C^\infty(M). \tag{7.37}$$

Its coordinate representation is identical to (6.18).

The tangent vector of a curve C is defined as the directional derivative along C. More explicitly, let C be given by the smooth parameter representation

$$Q: [a, b] \to M$$
$$\tau \mapsto Q(\tau). \tag{7.38}$$

Then the tangent vector of C at $Q(\tau)$, denoted by $\dot{Q}(\tau)$ or $\dot{Q}|_\tau$, is an element of $T_{Q(\tau)} M$ and acts on functions which are differentiable at $Q(\tau)$ according to

$$\dot{Q}|_\tau(f) = \frac{d}{d\tau}(f \circ Q)(\tau). \tag{7.39}$$

If $(\mathcal{U}, \alpha)$ is a chart for M with coordinates $x^1, x^2, \ldots, x^n$ and $Q(\tau) \in \mathcal{U}$, we get as the coordinate representation of C

$$Q^i(\tau) = \alpha^i(Q(\tau)), \tag{7.40}$$

and we can decompose $\dot{Q}(\tau)$ with respect to the corresponding holonomic basis (cf. (2.12)):

$$\dot{Q}(\tau) = \sum_{i=1}^{n} \frac{dQ^i}{d\tau} \frac{\partial}{\partial x^i}\bigg|_{Q(\tau)}. \tag{7.41}$$

7.6 Frames

An n-frame (or frame for short) is a set of n vector fields $b_1, b_2, \ldots, b_n$ defined on some subset of an n-dimensional manifold M such that they are linearly independent at each point of their domain of definition (cf. section 2.2). Consequently, the tangent vectors $b_1(x), b_2(x), \ldots, b_n(x)$ form a basis of $T_x M$.

A coordinate system $(\mathscr{U}, \alpha)$ with coordinates $x^1, x^2, \ldots, x^n$ determines an associated holonomic n-frame $\partial/\partial x^1, \partial/\partial x^2, \ldots, \partial/\partial x^n$ on $\mathscr{U}$. However, in general there is no frame defined on all of M. If such a global frame does exist, M is called parallelizable. Note that Cartesian products of parallelizable manifolds are parallelizable.

Since open subsets of $\mathbb{R}^n$ may be covered by one coordinate neighbourhood, the corresponding holonomic frame is globally defined. Consequently, open subsets of $\mathbb{R}^n$ are parallelizable manifolds. Furthermore, all Lie groups are parallelizable (see chapter 8). Out of the n-spheres, only S^1, S^3, and S^7 are parallelizable.

On the other hand, we have the famous hedgehog theorem: A smoothly combed hedgehog has at least one point of baldness. In mathematical terms: On an even-dimensional sphere every smooth vector field has at least one zero. So one cannot even find a single vector field which is everywhere linearly independent.

As we shall see, parallelizable manifolds are particularly convenient. For instance, they always possess a spin structure (see chapter 11). Furthermore, in general relativity the restriction to parallelizable manifolds is not too severe: Usually one takes as spacetime manifold the Cartesian product of $\mathbb{R}$ and a three-dimensional manifold. But a three-dimensional manifold is parallelizable, if it is orientable (Stiefel 1936). (The concept of orientability will be explained shortly.) So, in the orientable case, one automatically gets parallelizable spacetimes.

7.7 The tangent mapping

For a smooth map

$$F: M \to N \tag{7.42}$$

from the manifold M to the manifold N we want to define the corresponding tangent mapping

$$T_x F: T_x M \to T_{F(x)} N. \tag{7.43}$$

Introducing coordinates $x^1, x^2, \ldots, x^n$ on a neighbourhood of $x \in M$ and coordinates $y^1, y^2, \ldots, y^p$ on a neighbourhood of $F(x)$, we reduce the case at hand to the situation considered in section 2.3: The map F is represented by

$$y^j = F^j(x^1, x^2, \ldots, x^n), \quad j = 1, 2, \ldots, p, \tag{7.44}$$

and for

$$v = \sum_{i=1}^{n} v^i \frac{\partial}{\partial x^i}(x) \in T_x M \tag{7.45}$$

we define (cf. (2.21)):

$$T_x F(v) = \sum_{j=1}^{p} \sum_{i=1}^{n} v^i \frac{\partial F^j}{\partial x^i} \frac{\partial}{\partial y^j}\bigg|_{F(x)}. \tag{7.46}$$

The motivation with the help of curves given in section 2.3 shows that (7.46) is independent of the chosen coordinates. However, it is also easy to check the coordinate independence explicitly. For the composition of two maps we have (cf. (2.24)):

$$T_x(G \circ F) = T_{F(x)} G \circ T_x F. \tag{7.47}$$

For later purposes we note that the tangent space of the Cartesian product of manifolds M and N at $(x, y) \in M \times N$ can be identified with $T_x M \times T_y N$ (or the direct sum of $T_x M$ and $T_y N$). Let $\pi_1(\pi_2)$ be the projection onto the first (second) factor of $M \times N$. Then we identify $v \in T_{(x,y)} M \times N$ with $(v_M, v_N) \in T_x M \times T_y N$ where $v_M = T_{(x,y)} \pi_1(v)$ and $v_N = T_{(x,y)} \pi_2(v)$.

7.8 Differential forms on manifolds

Generalizing the definition given in section 2.4, we have: A p-form on a manifold M maps $x \in M$ onto $\varphi_x \in \Lambda^p T_x M$. The x-dependence of φ is required to be smooth, i.e. for $v_1, v_2, \ldots, v_p \in \text{vect}(M)$ the real-valued function f with

$$f(x) := \varphi_x(v_1(x), v_2(x), \ldots, v_p(x)) \tag{7.48}$$

must lie in $C^\infty(M)$.

The real vector space of p-forms on M is denoted by $\Lambda^p M$. We have $\Lambda^0 M = C^\infty(M)$ and put

$$\Lambda M := \bigoplus_{p=0}^{n} \Lambda^p M, \tag{7.49}$$

where $n = \dim M$. It should be obvious how to define differential forms with values in a real vector space W different from $\mathbb{R}$. The set of all W-valued p-forms on M is denoted by $\Lambda^p(M, W)$. We shall also consider forms whose domain is only a subset of M.

Once more, the product of a function and a differential form, the wedge product of two forms and the inner derivative of a form with respect to a vector field are defined pointwise.

A set of n 1-forms which are linearly independent at each point where they are defined is again called an n-frame (dual n-frame, if we want to distinguish it from an n-frame consisting of vector fields). Obviously, the discussion of section 7.6 is valid for these frames, too.

In terms of coordinates we have the same formulas as in chapter 2. Let $x^1, x^2, \ldots, x^n$ be coordinates on some open subset $\mathcal{U}$ of M and $\partial/\partial x^1, \partial/\partial x^2, \ldots, \partial/\partial x^n$ the associated holonomic frame. Denoting the corresponding dual holonomic frame by $dx^1, dx^2, \ldots, dx^n$ we can express $\varphi \in \Lambda^p M$ on $\mathcal{U}$ as

$$\varphi = \frac{1}{p!} \sum_{i_1,\ldots,i_p = 1}^{n} \varphi_{i_1 \cdots i_p} \, dx^{i_1} \wedge \cdots \wedge dx^{i_p} \tag{7.50}$$

with smooth functions $\varphi_{i_1 \cdots i_p}$.

Under a change of coordinates the dual holonomic frame transforms according to (2.29), and from

$$\varphi = \frac{1}{p!} \sum_{i_1,\ldots,i_p = 1}^{n} \varphi_{i_1 \cdots i_p} \, dx^{i_1} \wedge \cdots \wedge dx^{i_p}$$

$$= \frac{1}{p!} \sum_{j_1,\ldots,j_p = 1}^{n} \varphi'_{j_1 \cdots j_p} \, dy^{j_1} \wedge \cdots \wedge dy^{j_p} \tag{7.51}$$

we can read off the transformation behaviour of the coefficients $\varphi_{i_1 \cdots i_p}$:

$$\varphi'_{j_1 \cdots j_p} = \sum_{i_1,\ldots,i_p = 1}^{n} \frac{\partial x^{i_1}}{\partial y^{j_1}} \cdots \frac{\partial x^{i_p}}{\partial y^{j_p}} \varphi_{i_1 \cdots i_p}. \tag{7.52}$$

As in section 2.5 we introduce the pull back of differential forms by a smooth mapping

$$F: M \to N, \tag{7.53}$$

where M and N are manifolds. For $\varphi \in \Lambda^p N$ we define $F^*\varphi \in \Lambda^p M$ by

$$(F^*\varphi)_x(v_1, v_2, \ldots, v_p) := \varphi_{F(x)}(T_x F(v_1), T_x F(v_2), \ldots, T_x F(v_p)) \quad (7.54)$$

with $x \in M$, $v_i \in T_x M$ (cf. (2.37)). The pull back possesses the properties (a), (b), (c) of section 2.5 in this more general setting too, and in terms of coordinates it has the same representation as in chapter 2.

The exterior derivative d is also easily generalized to differential forms on manifolds. With the help of coordinates $x^1, x^2, \ldots, x^n$ we write $\varphi \in \Lambda^p M$ locally as

$$\varphi = \frac{1}{p!} \sum_{i_1, \ldots, i_p = 1}^{n} \varphi_{i_1 \cdots i_p}(x^1, \ldots, x^n) \, dx^{i_1} \wedge \cdots \wedge dx^{i_p} \quad (7.55)$$

and define $d\varphi \in \Lambda^{p+1} M$ according to (cf. (2.60))

$$d\varphi := \frac{1}{p!} \sum_{i, i_1, \ldots, i_p = 1}^{n} \frac{\partial}{\partial x^i} \varphi_{i_1 \cdots i_p}(x^1, \ldots, x^n) \, dx^i \wedge dx^{i_1} \wedge \cdots \wedge dx^{i_p}. \quad (7.56)$$

As we have shown in section 2.6, definition (7.56) is coordinate independent. In particular, this means that the forms given locally by (7.56) fit together so that $d\varphi$ is well defined on all M. Obviously, the operator $d: \Lambda M \to \Lambda M$ enjoys the properties (a) to (e) of the theorem in section 2.6.

Again, d commutes with the pull back:

$$F^* d\varphi = dF^* \varphi, \quad (7.57)$$

if $F: M \to N$ is a smooth mapping of manifolds and $\varphi \in \Lambda^p N$. Also the Lie derivative of differential forms is given by the same expression as before (cf. (6.34)):

$$L_v = i_v d + d i_v \quad (7.58)$$

for $v \in \text{vect}(M)$. The flow of the vector field v introduced in exactly the same manner as in section 6.1 can be used to express the Lie derivative (7.58) by an equation similar to (6.12). Closed and exact forms, the de Rham cohomology groups, the Betti numbers, and the Euler characteristic of a manifold M are defined and denoted as in section 2.7 with $\mathcal{U}$ replaced by M.

In order to formulate a generalization of the Poincaré lemma of section 2.7 we need the notion of a contractible manifold. A manifold M is called contractible onto a point $x_0 \in M$ if the identity map on M may be deformed smoothly into the constant map x_0; i.e. if there is a smooth map

$$F: [0, 1] \times M \to M \quad (7.59)$$

with

$$F(0, x) = x, \quad F(1, x) = x_0 \tag{7.60}$$

for all $x \in M$. Upon interpreting the first argument of F as the time t, the point x moves along the curve in M given by $F(t, x)$ from $F(0, x) = x$ to $F(1, x) = x_0$ during the contraction. For example, a star-shaped open subset of $\mathbb{R}^n$ is contractible, whereas $\mathbb{R}^n - \{0\}$ is not. Now the Poincaré lemma takes the following form.

THEOREM

On a contractible manifold M, a differential form $\varphi \in \Lambda M$ is closed if and only if it is exact.

So, if M is contractible, we have

$$H^p(M) = \{0\} \quad \text{for } p > 0. \tag{7.61}$$

For later purposes we need one more definition. A manifold M is called simply connected if the following two conditions are met:

(a) Any two points of M may be joined by a smooth curve.
(b) Any closed curve in M may be smoothly deformed into a constant curve.

Condition (b) means: For any closed curve given by a parameter representation

$$Q: [0, 1] \to M, \quad Q(0) = Q(1), \tag{7.62}$$

there is a smooth map

$$F: [0, 1] \times [0, 1] \to M \tag{7.63}$$

such that

$$
\begin{aligned}
F(0, \tau) &= Q(\tau), && 0 \leqslant \tau \leqslant 1, \\
F(1, \tau) &= Q(0), && 0 \leqslant \tau \leqslant 1, \\
F(t, 0) &= F(t, 1) = Q(0), && 0 \leqslant t \leqslant 1.
\end{aligned}
\tag{7.64}
$$

A contractible manifold is simply connected.

7.9 Orientable manifolds

An orientation of an n-dimensional manifold M means a coherent orientation of all the tangent spaces $T_x M$. It can be given, for example, by a nowhere vanishing n-form $\omega \in \Lambda^n M$ (strictly speaking, an equivalence class of n-forms, see section 2.8). A manifold equipped with an orientation is called an oriented manifold.

An orientation does not exist on all manifolds. The Möbius strip, for instance, does not admit an orientation. If it does exist, the manifold is called orientable. Note that a connected orientable manifold possesses exactly two orientations. Any parallelizable manifold is orientable: If $\beta^1, \beta^2, \ldots, \beta^n$ denotes a global n-frame in cotangent space, the n-form $\omega = \beta^1 \wedge \beta^2 \wedge \cdots \wedge \beta^n$ defines an orientation.

On an oriented manifold (orientation given by $\omega \in \Lambda^n M$) a chart $(\mathcal{U}, \alpha)$ with coordinates $x^1, x^2, \ldots, x^n$ is called oriented, if the associated holonomic frame is oriented, i.e. if

$$\omega_x\left(\frac{\partial}{\partial x^1}(x), \ldots, \frac{\partial}{\partial x^n}(x)\right) > 0 \qquad (7.65)$$

for all $x \in \mathcal{U}$ or equivalently, if

$$dx^1 \wedge dx^2 \wedge \cdots \wedge dx^n = f\omega, \qquad (7.66)$$

where f is an everywhere positive function on $\mathcal{U}$. An atlas consisting of oriented charts only is called oriented. Since

$$dy^1 \wedge dy^2 \wedge \cdots \wedge dy^n = \det\left(\frac{\partial y^j}{\partial x^i}\right) dx^1 \wedge dx^2 \wedge \cdots \wedge dx^n \qquad (7.67)$$

for two coordinate systems $x^1, x^2, \ldots, x^n$ and $y^1, y^2, \ldots, y^n$, we see: In the case of an oriented atlas all coordinate transformations have positive Jacobian. Conversely, one can define an orientation by exhibiting an atlas such that all changes of coordinates have positive Jacobian, i.e. by gluing together local coordinate systems in an orientation-preserving manner.

7.10 Manifolds with boundary

An n-dimensional manifold was defined to be a space which locally 'looks like an open subset of $\mathbb{R}^n$'. Replacing $\mathbb{R}^4$ by the half space

$$\mathbb{H}^n := \{(x^1, x^2, \ldots, x^n) \in \mathbb{R}^n, x^n \geq 0\} \qquad (7.68)$$

we arrive at the notion of a manifold with boundary. So, in this case an atlas consists of charts $(\mathcal{U}_r, \alpha_r)$ such that α_r is a homeomorphism of $\mathcal{U}_r$ onto an open subset of $\mathbb{H}^n$ (see fig. 7.4). With obvious modifications, everything discussed so far in this chapter remains valid in this more general situation.

The boundary of M, denoted by ∂M, is the set of all points which are

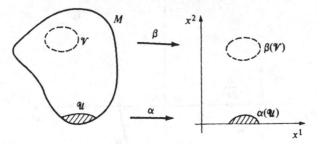

Fig. 7.4. Manifold with boundary.

mapped onto points in

$$\partial\mathbb{H}^n := \{(x^1, x^2, \ldots, x^n) \in \mathbb{R}^n \mid x^n = 0\} \tag{7.69}$$

by the coordinate maps α_r; i.e.

$$\partial M = \bigcup_r \alpha_r^{-1}(\partial\mathbb{H}^n). \tag{7.70}$$

For a manifold in the previous sense, i.e. a manifold without boundary, we have $\partial M = \varnothing$. One can prove that the above definition of ∂M is indeed independent of the coordinates used. Furthermore, ∂M is an $(n-1)$-dimensional manifold without boundary:

$$\partial\partial M = \varnothing. \tag{7.71}$$

A simple example of a manifold with boundary is provided by the closed ball

$$B^n := \left\{(x^1, x^2, \ldots, x^n) \in \mathbb{R}^n \,\middle|\, \sum_{i=1}^n (x^i)^2 \leqslant 1\right\}, \tag{7.72}$$

whose boundary is the $(n-1)$-sphere:

$$\partial B^n = S^{n-1}. \tag{7.73}$$

Let M be a manifold with boundary and $\mathcal{U} \subset M$ a coordinate neighbourhood, homeomorphic to an open subset of $\mathbb{H}^n$, such that $\mathcal{U} \cap \partial M \neq \varnothing$. Denote the corresponding coordinates by $x^1, x^2, \ldots, x^n$. The tangent space $T_x \partial M$ may be considered as a subspace of $T_x M$, because $T_x \partial M$ is spanned by

$$\frac{\partial}{\partial x^1}(x), \frac{\partial}{\partial x^2}(x), \ldots, \frac{\partial}{\partial x^{n-1}}(x),$$

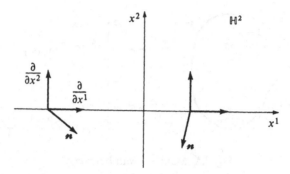

Fig. 7.5. Outward normal vectors on $\partial\mathbb{H}^2$.

for $x \in \mathscr{U} \cap \partial M$. On the other hand, a tangent vector

$$\sum_{i=1}^{n} t^i \frac{\partial}{\partial x^i}(x) \in T_x M, \quad x \in \mathscr{U} \cap \partial M, \tag{7.74}$$

with $t^n < 0$ is called an outward normal vector. The geometrical meaning of this definition becomes clear in the case of the half space $\mathbb{H}^n$, the prototype of a manifold with boundary (see fig. 7.5). Note that here the word 'normal' does not refer to any kind of orthogonality. (We do not have a metric yet!) It just indicates that the vector (7.74) is not tangent to ∂M. One can prove that there exists an outward normal vector field n on ∂M, i.e a smooth map which assigns to each $x \in \partial M$ an outward normal vector $n(x) \in T_x M$.

If M is oriented, we can use such an outward normal vector field n in order to define an induced orientation of ∂M (cf. section 2.10). Let the orientation of M be given by $\omega \in \Lambda^n M$. For $x \in \partial M$ and $v_1, v_2, \dots, v_{n-1} \in T_x \partial M \subset T_x M$ put

$$\tilde{\omega}_x(v_1, v_2, \dots, v_{n-1}) = \omega_x(n(x), v_1, v_2, \dots, v_{n-1}). \tag{7.75}$$

Then $\tilde{\omega} \in \Lambda^{n-1} \partial M$ defines an orientation of ∂M, which can be shown to be independent of the choice of n. Whenever M is oriented, ∂M is considered to be oriented in this way. In terms of coordinates we can say: $y^1, y^2, \dots, y^{n-1}$ is an oriented coordinate system for ∂M, provided:

$$n(x), \frac{\partial}{\partial y^1}(x), \frac{\partial}{\partial y^2}(x), \dots, \frac{\partial}{\partial y^{n-1}}(x)$$

is an oriented basis of $T_x M$, $x \in \partial M$.

For example, take $\mathbb{H}^n$ oriented by means of $dx^1 \wedge dx^2 \wedge \cdots \wedge dx^n$, where $x^1, x^2, \ldots, x^n$ are Cartesian coordinates for $\mathbb{R}^n$. As an outward normal vector field one may choose

$$n(x) = -\frac{\partial}{\partial x^n}(x), \quad x \in \partial \mathbb{H}^n. \tag{7.76}$$

Then the induced orientation of $\partial \mathbb{H}^n$ is described by $(-1)^n dx^1 \wedge dx^2 \wedge \cdots \wedge dx^{n-1} \in \Lambda^{n-1} \partial \mathbb{H}^n$.

In the following, we shall equip the sphere S^{n-1} with the orientation that it inherits from $\mathbb{R}^n$ via the representation (7.73).

7.11 The partition of unity

In order to prove the existence of a global object which, to begin with, can be defined only locally, one frequently uses a so-called partition of unity.

Let M be a manifold and $\{\mathcal{U}_r\}$ a locally finite open covering of M; i.e. each $x \in M$ possesses a neighbourhood $\mathcal{U}$ such that $\mathcal{U} \cap \mathcal{U}_r \neq \varnothing$ only for a finite number of rs. A partition of unity subordinate to this covering is a family $\{h_r\}$ of smooth functions $h_r \in C^\infty(M)$ with the properties:

(a) $h_r(x) \geq 0$ for all $x \in M$;
(b) $\text{supp}(h_r) \subset \mathcal{U}_r$ for all r;
(c) $\sum_r h_r(x) = 1$ for all $x \in M$.

The sum in (c) contains only a finite number of nonvanishing terms due to (b) and the local finiteness of the covering $\{\mathcal{U}_r\}$. So it is well defined.

THEOREM

There exists a partition of unity subordinate to any locally finite open covering of M.

For later applications it is useful to note that every manifold has an atlas $\{(\mathcal{U}_r, \alpha_r)\}$ such that $\{\mathcal{U}_r\}$ is locally finite.

As an example consider the manifold $M = \mathbb{R}$ together with the locally finite open covering $\{(r-2, r+2)\}, r \in \mathbb{Z}$. Define $\vartheta \in C^\infty(\mathbb{R})$ by

$$\vartheta(x) := \begin{cases} \exp(-(1-x^2)^{-1}), & |x| < 1, \\ 0, & |x| \geq 1. \end{cases} \tag{7.77}$$

Then it is easily checked that the functions

$$h_r(x) := \frac{\vartheta(x-r)}{\sum\limits_{s=-\infty}^{\infty} \vartheta(x-s)} \tag{7.78}$$

form a partition of unity subordinate to the given covering.

7.12 Integration

Let M be an oriented n-dimensional manifold (with or without boundary), $\{(\mathcal{U}_r, \alpha_r)\}$ an oriented atlas such that $\{\mathcal{U}_r\}$ is a locally finite open covering of M, and $\{h_r\}$ a partition of unity subordinate to this covering. Call the n-form which is to be integrated φ. We start with the special case of a domain of integration K which is contained in a single coordinate neighbourhood $\mathcal{U}_r$. Moreover, of course, K has to be 'sufficiently regular'. In terms of oriented coordinates $x^1, x^2, \ldots, x^n$ on $\mathcal{U}_r$ we can write

$$\varphi = f \, dx^1 \wedge dx^2 \wedge \cdots \wedge dx^n \tag{7.79}$$

with a smooth function f on $\mathcal{U}_r$. Remembering that $x^1, x^2, \ldots, x^n$ may be interpreted as the Cartesian coordinates on $\alpha_r(\mathcal{U}_r)$ we shall consider them as an oriented coordinate system also for $\alpha_r(\mathcal{U}_r)$. The integral of φ over K is then defined to be the integral of the pull back of φ by the inverse coordinate map α_r^{-1} over $\alpha_r(K) \subset \alpha_r(\mathcal{U}_r)$:

$$\int_K \varphi := \int_{\alpha_r(K)} (\alpha_r^{-1})^* \varphi = \int \int \cdots \int_{\alpha_r(K)} f(x^1, x^2, \ldots, x^n) \, dx^1 \, dx^2 \cdots dx^n. \tag{7.80}$$

Note how the existence of coordinates enables us to reduce the integration on a manifold to an integration in $\mathbb{R}^n$ which was discussed in section 2.8.

If K is not contained in a single $\mathcal{U}_r$, we use the partition of unity to express $\int_K \varphi$ in terms of integrals already defined:

$$\int_K \varphi := \sum_r \int_{K \cap \mathcal{U}_r} h_r \varphi. \tag{7.81}$$

The introduction of a partition of unity is necessary because in general $(K \cap \mathcal{U}_r) \cap (K \cap \mathcal{U}_s) \neq \varnothing$ even for $r \neq s$. Usually, of course, some further technical assumptions have to be made to ensure the existence of the integral, e.g. compactness of K or $\mathrm{supp}(\varphi)$. One can prove that the above definition is independent of the choices involved, in particular it does not depend on the partition of unity which is used.

The integral (7.81) inherits all desirable properties from its ancestor in $\mathbb{R}^n$.

For example, let M and N be n-dimensional manifolds with orientations given by $\omega \in \Lambda^n M$ and $\omega' \in \Lambda^n N$, respectively, φ an n-form on N and $K \subset M$. If

$$F: M \to N \tag{7.82}$$

is an orientation-preserving diffeomorphism, i.e. $F^* \omega'$ defines the same orientation of M as ω, we have

$$\int_K F^* \varphi = \int_{F(K)} \varphi. \tag{7.83}$$

Stokes' theorem now takes the following form:

$$\int_M d\varphi = \int_{\partial M} \varphi, \tag{7.84}$$

where M is an n-dimensional manifold with boundary, which can be empty, and $\varphi \in \Lambda^{n-1} M$.

7.13 Metric

Generalizing the discussion of chapter 3 we define a (pseudo-) metric on an n-dimensional manifold M to be given by a nondegenerate symmetric bilinear form g_x on each tangent space $T_x M$, which depends smoothly on x; i.e. for $v_1, v_2 \in \text{vect}(M)$ the function

$$x \mapsto g_x(v_1(x), v_2(x)) \tag{7.85}$$

lies in $C^\infty(M)$. A manifold equipped with a metric is called a Riemannian manifold.

A positive-definite metric exists on any manifold: Let $\{\mathcal{U}_r\}$ be a locally finite covering of M by coordinate neighbourhoods and $\{h_r\}$ a partition of unity subordinate to that covering. On each $\mathcal{U}_r$ one easily defines a positive-definite metric g^r, e.g. by

$$g^r_x\left(\frac{\partial}{\partial x^i}(x), \frac{\partial}{\partial x^j}(x)\right) = \delta_{ij}, \tag{7.86}$$

where $x^1, x^2, \ldots, x^n$ are coordinates on $\mathcal{U}_r$. Then

$$g := \sum_r h_r g^r \tag{7.87}$$

is a positive-definite metric on M, because the sum of two positive-definite bilinear forms is again positive-definite.

On the other hand, a metric of Minkowski signature $(+, -, -, -)$ does not exist on all four-dimensional manifolds. But every noncompact manifold admits such a metric, as well as every parallelizable manifold. If M is compact and connected, it possesses a metric of Minkowski signature provided its Euler characteristic is zero. Conversely, if a compact connected orientable manifold admits such a metric, then its Euler characteristic vanishes. Note that compact spacetime manifolds are anyhow unreasonable from the physical point of view, because they have closed timelike curves mixing up past and future.

For an n-dimensional Riemannian manifold M we can define isometries and the canonical isomorphism J of $T_x M$ and $(T_x M)^*$ by complete analogy with chapter 3. If M is oriented, we have a volume form (or volume element) given in terms of an oriented orthonormal frame e^i by $e^1 \wedge e^2 \wedge \cdots \wedge e^n$, and we get the Hodge star

$$*: \Lambda^p M \to \Lambda^{n-p} M$$
$$\varphi \mapsto *\varphi \tag{7.88}$$

by pointwise application of the $*$ operation introduced in section 3.3. The definitions of the coderivative, the Laplacian, and the d'Alembert operator are the obvious generalizations of those in chapter 3. Furthermore, the expressions for the gradient, divergence, and curl in terms of J, d, $*$, and δ (cf. (3.48)–(3.50)) also remain valid in the present context.

7.14 Gauge theories

Upon replacing the open subset $\mathcal{U} \subset \mathbb{R}^4$ by an orientable four-dimensional manifold M with a metric of Minkowski signature the results of chapter 4 are unchanged, provided one considers only trivial bundles over M. The same remark applies to the discussions of chapter 5 if M is parallelizable, i.e. if the tangent bundle of M is trivial. As mentioned at the end of section 7.6, the assumption of parallelizability is not too restrictive in general relativity. The generalization to nontrivial bundles (nonparallelizable manifolds) as well as the whole bundle formalism will be given in chapter 9.

Problems

7.1 Show that the definition of differentiability of maps between two manifolds is independent of the coordinates chosen.

7.2 Prove the theorem in section 7.4.

7.3 Show that the differentiable structure of S^2 as a submanifold of $\mathbb{R}^3$

coincides with the differentiable structure given in section 7.1. Hint: Use polar coordinates in $\mathbb{R}^3$.

7.4 Verify that the Lie bracket of two vector fields on a manifold is represented in terms of coordinates by (6.18).

7.5 Prove the coordinate independence of the definition (7.46) of the tangent mapping explicitly.

8

Lie groups

In this chapter we consider Lie groups, Lie algebras, and their representations from a geometrical point of view: A Lie group is a particular
manifold. So far we have described spacetime by a manifold M. On the
other hand, a gauge group ^MG is the set of functions from spacetime into a
Lie group G, i.e. $^MG \subset M \times G$. Therefore by considering also G as a
manifold we arrive at a more homogeneous formulation which is naturally
generalized to fibre bundles in the next chapter.

8.1 Definition and examples

A (real, finite-dimensional) Lie group is a set G which is at the same time a
differentiable manifold and a group such that its group structure is
differentiable. This means that the two mappings

$$G \times G \to G$$

$$(g, g') \mapsto gg'$$

and

$$G \to G$$

$$g \mapsto g^{-1}$$

are differentiable.

The set G inherits from its manifold structure properties like dimensionality, compactness, connectedness, simple connectedness etc., and from its
group structure adjectives like abelian, simple, nilpotent, etc. A Lie
subgroup is a subgroup which is also a submanifold. We denote by e the
neutral element of G and by G^e the connected component of e. The set G^e is a
Lie subgroup with the same dimension as G.

There are only two different connected one-dimensional Lie groups: First
$\mathbb{R}$ with its additive group structure. It is noncompact and simply connected.
The second is $S^1 := \{z \in \mathbb{C}, |z| = 1\}$ with its multiplicative group
structure, also denoted by $U(1)$ or $SO(2)$. It is compact and not simply

connected. Both groups are abelian. More generally, any connected abelian Lie group is a Cartesian product of copies of the two above groups.

A more interesting example is the general linear group GL_n, the set of real $n \times n$ matrices with nonvanishing determinant. It consists of two open subsets of $\mathbb{R}^{n^2}$, the matrices of negative and positive determinant, respectively. GL_n is of dimension n^2, not compact, not connected, not simply connected, nonabelian ($n \geqslant 2$). The connected component of the neutral element is GL_n^+.

The special orthogonal group $SO(n)$ is the subset of GL_n consisting of orthogonal matrices γ,

$$\gamma^T \gamma = \mathbb{1}, \tag{8.1}$$

with unit determinant. The group $SO(n)$ is a manifold of dimension $\frac{1}{2}n(n-1)$, it is the maximal compact Lie subgroup of GL_n^+. It is connected but not simply connected. Another class of examples are the special unitary groups $SU(n)$, consisting of all $n \times n$ matrices U with complex coefficients which are unitary,

$$U^+ U = \mathbb{1}, \tag{8.2}$$

and have determinant one. We stress that $SU(n)$ is a real Lie group, a Lie subgroup of GL_{2n}. It is of dimension $n^2 - 1$, compact, and simply connected.

As a manifold, $SU(2)$ is S^3: We write an $SU(2)$-element as

$$U = \begin{pmatrix} a & b \\ c & d \end{pmatrix}, \quad a, b, c, d \in \mathbb{C}. \tag{8.3}$$

Unitarity means that column vectors of U are orthonormal,

$$\bar{a}a + \bar{c}c = 1, \tag{8.4}$$

$$\bar{b}b + \bar{d}d = 1, \tag{8.5}$$

$$\bar{a}b + \bar{c}d = 0, \tag{8.6}$$

and the condition of unit determinant reads

$$ad - cb = 1. \tag{8.7}$$

These four quadratic equations are equivalent to

$$\bar{a}a + \bar{c}c = 1, \tag{8.8}$$

$$b = -\bar{c}, \tag{8.9}$$

$$d = \bar{a}. \tag{8.10}$$

Therefore, we get the most general $SU(2)$-matrix in the form

$$U = \begin{pmatrix} x_1 + ix_2 & -x_3 + ix_4 \\ x_3 + ix_4 & x_1 - ix_2 \end{pmatrix}, \quad x_i \in \mathbb{R}, \tag{8.11}$$

with

$$\sum_{i=1}^{4} x_i^2 = 1. \tag{8.12}$$

So we have constructed a diffeomorphism from $SU(2)$ to S^3.

8.2 Representations of a Lie group

Let G be a Lie group, M a manifold. A representation of G on M is a differentiable mapping

$$G \times M \to M$$

$$(g, x) \mapsto \rho_g x$$

such that for each group element g

$$\rho_g : M \to M$$

is a diffeomorphism of M satisfying

$$\rho_e = \mathrm{id}_M, \tag{8.13}$$

$$\rho_g \circ \rho_{g'} = \rho_{gg'}. \tag{8.14}$$

In other words, a representation is a group homeomorphism from G to Diff (M).

Examples are the representations of the rotation group $SO(3)$ on our three-dimensional space or on the 2-sphere. More generally, the isometries of a Riemannian manifold form a Lie group which is represented on this manifold.

For notational convenience we sometimes write $\rho(g)$ instead of ρ_g. Some authors reserve the word representation for the linear case to be defined below. In general a representation is also called realization, action, or left action. A right action is by definition the same construction as above, with (8.14) replaced by

$$\rho_g \circ \rho_{g'} = \rho_{g'g}. \tag{8.15}$$

Note that a right action ρ_g is not a representation, but the map: $g \mapsto \rho_{g^{-1}}$ is a representation and conversely starting from a representation ρ, $\rho_{g^{-1}}$ always defines a right action. Sometimes left and right actions are abbreviated

$$\rho_g x =: gx, \quad \text{left action,} \tag{8.16}$$

$$\rho_g x =: xg, \quad \text{right action.} \tag{8.17}$$

We introduce some vocabulary related to a representation ρ of G on M. For any point x in M we define the map

$$\text{bit}_x : G \to M$$
$$g \mapsto \text{bit}_x g := \rho_{g^{-1}} x, \tag{8.18}$$

in particular, $\text{bit}_x e = x$. By definition the orbit of x is the image of bit_x:

$$\text{orbit}(x) := \text{bit}_x(G); \tag{8.19}$$

i.e. the orbit of x consists of all points in M which can be reached from x by a group transformation.

The representation is called transitive if M consists of only one orbit. This means that any two points x and y in M are related by some group element g:

$$y = \rho_g x. \tag{8.20}$$

A representation is called faithful (or effective) if the homeomorphism $g \mapsto \rho_g$ is injective. Note that this homeomorphism is never surjective since $\text{Diff}(M)$ is infinite-dimensional. If bit_x is injective for some x in M then the representation is faithful.

A representation is called free if only ρ_e has a fixed point: If for some x in M $\rho_g x = x$, then g must be the neutral element. A representation is free if and only if bit_x is injective for all x in M. Of course, free implies faithful.

Finally, for a given point x in M we define its isotropy group

$$I(x) := \{ g \in G \,|\, \rho_g x = x \} \tag{8.21}$$

as the subgroup of G consisting of those elements that leave x fixed. A representation is free if and only if all isotropy groups are trivial: $I(x) = \{e\}$.

The representation of $SO(3)$ on $\mathbb{R}^3$ is not transitive, its orbits are the origin and the spheres of different radii, it is faithful and not free. Its isotropy groups are $SO(2)$ for all points except the origin where the isotropy group is $SO(3)$ itself.

The representation of $SO(3)$ on S^2 is transitive, faithful, and not free. All its isotropy groups are $SO(2)$. This last fact is sometimes phrased by physicists: 'S^2 is a nonlinear realization of $SO(3)$ down to $SO(2)$'.

We now discuss important examples of representations of a Lie group on itself: $M = G$. We first define the left translations L_g:

$$G \times G \to G$$

$$(g, h) \mapsto L_g h := gh, \qquad (8.22)$$

where g acts on h by left multiplication. This representation is transitive and free.

Another representation of G on itself is given by the inner automorphisms:

$$G \times G \to G$$

$$(g, h) \mapsto \mathrm{aut}_g h := ghg^{-1}. \qquad (8.23)$$

Note that $I(e) = G$.

Finally we define the right action of right translations R_g:

$$R_g h := hg. \qquad (8.24)$$

They commute with the left translations and

$$\mathrm{aut}_g = L_g \circ R_{g^{-1}} = R_{g^{-1}} \circ L_g. \qquad (8.25)$$

A representation ρ is called linear if the manifold it operates on is a vector space V and if all diffeomorphisms ρ_g are linear. In other words, a linear representation is a group homeomorphism

$$G \to GL(V) \subset \mathrm{Diff}(V).$$

Of course no linear representation is transitive and we call a linear representation irreducible if V contains no invariant subspace W,

$$\rho_g W \subset W \quad \text{for all} \quad g \in G,$$

which is nontrivial: $W \neq V, W \neq \{0\}$. The fundamental representation of $SO(3)$ on $\mathbb{R}^3$ is linear and irreducible.

A representation is called affine if it operates on a vector space V and if all ρ_g are affine maps. Recall that a map from V into itself is affine if it is the sum of a linear and a constant map. For example the defining representation of the group of translations in $\mathbb{R}^3$ is affine.

Any linear or affine representation on a vector space V can be given a nonlinear appearance by means of a fixed nonlinear diffeomorphism F from V onto itself (sometimes called coordinate transformation):

$$(g, v) \mapsto \rho'_g v := F \rho_g F^{-1} v. \qquad (8.26)$$

Conversely, given a representation defined on a vector space, it is in general not easy to decide whether it is a linear or affine representation disguised by means of some diffeomorphism F.

In this chapter we restrict ourselves to finite-dimensional Lie groups and finite-dimensional representations. Analogous definitions are used for infinite-dimensional representations and infinite-dimensional groups, e.g. gauge groups. In this sense the frames and the components of a given metric with respect to a frame carry truly nonlinear representations of the gauge group $^{\#}GL_4$ defined by equations (5.1), (5.2), respectively. On the space of connections we have an affine representation given by equation (5.3). Similarly, curvature and torsion give rise to linear representations of $^{\#}GL_4$ (cf. (5.10), (5.13)).

8.3 From a Lie group to its Lie algebra

A vector field A on a Lie group G is called (left) invariant if it is invariant under all left translations:

$$T_h L_g(A(h)) = A(gh) \quad \text{for all} \quad g \in G. \tag{8.27}$$

For instance, the invariant vector fields on the abelian Lie group $\mathbb{R}^n$ are the constant vector fields.

A linear combination of invariant vector fields is invariant, because $T_h L_g$ is a linear mapping. Also the Lie bracket of two invariant vector fields is invariant. Therefore the invariant vector fields form a Lie subalgebra of vect(G). We denote this subalgebra by g, Lie(G), or in the case of the matrix groups GL_n, $SO(n)$, $SU(n)$, ... also by gl_n, $so(n)$, $su(n)$, etc. g is called the Lie algebra of the Lie group G. It is finite-dimensional and

$$\dim \mathfrak{g} = \dim G. \tag{8.28}$$

Indeed each invariant vector field A is uniquely determined by its value $A(e) \in T_e G$ at the neutral element. As L_g is a transitive representation, we have

$$A(g) = T_e L_g A(e). \tag{8.29}$$

If d is the dimension of G and $A_1(e), \ldots, A_d(e)$ a basis of $T_e G$, then the corresponding invariant vector fields $A_1, \ldots, A_d$ are a basis of g. Moreover, these vector fields are linearly independent in each point $g \in G$, thus forming a global frame on G. Therefore any Lie group is parallelizable. The Lie algebra describes the Lie group G only locally around the neutral element, in particular

$$\text{Lie}(G^e) = \text{Lie}(G). \tag{8.30}$$

We now discuss in detail the example GL_n. Note that most Lie groups have faithful linear representations and are thereby subgroups of some GL_n. Being an open subset of $\mathbb{R}^{n^2}$, GL_n has global coordinates. Let g be a GL_n-element. Its standard coordinates are $x^{ij}(g) \in \mathbb{R}$, the element in the ith row, jth column of g. In particular

$$x^{ij}(e) = \delta^i{}_j. \tag{8.31}$$

A tangent vector $A(e) \in T_e G$ can be written in these coordinates

$$A(e) = \sum_{i_1, i_2 = 1}^{n} a^{i_1 i_2} \frac{\partial}{\partial x^{i_1 i_2}}\bigg|_e. \tag{8.32}$$

The numbers $a^{i_1 i_2}$ form a constant $n \times n$ matrix. In order to compute the invariant vector field $A(g)$, we have to express the left translation L_g for fixed g in the coordinates $x^{j_1 j_2}$ (cf. (2.16)):

$$x^{j_1 j_2}(L_g h) = \sum_k x^{j_1 k}(g) x^{k j_2}(h). \tag{8.33}$$

Using the expression of the tangent map in coordinates, equation (2.21), we get:

$$\begin{aligned} A(g) &= T_e L_g A(e) \\ &= \sum_{\substack{i_1, i_2, \\ j_1, j_2, k}} a^{i_1 i_2} \frac{\partial}{\partial x^{i_1 i_2}} (x^{j_1 k}(g) x^{k j_2}) \frac{\partial}{\partial x^{j_1 j_2}}\bigg|_g \\ &= \sum_{\substack{i_1, i_2, \\ j_1, j_2, k}} a^{i_1 i_2} x^{j_1 k}(g) \delta^k{}_{i_1} \delta^{j_2}{}_{i_2} \frac{\partial}{\partial x^{j_1 j_2}}\bigg|_g \\ &= \sum_{i_1, i_2, k} x^{i_1 k}(g) a^{k i_2} \frac{\partial}{\partial x^{i_1 i_2}}\bigg|_g. \end{aligned} \tag{8.34}$$

In the literature the last expression is often abbreviated by $gA(e)$ where $A(e)$ is identified with the $n \times n$ matrix (a^{ij}) and the product is matrix multiplication.

Let us calculate the bracket of two invariant vector fields A and B in the coordinates x^{ij} (cf. 6.18)):

$$[A, B] = \sum_{\substack{i_1, i_2, \\ j_1, j_2, \\ l, k}} x^{i_1 k} a^{k i_2} \frac{\partial}{\partial x^{i_1 i_2}} (x^{j_1 l} b^{l j_2}) \frac{\partial}{\partial x^{j_1 j_2}} - (a \leftrightarrow b)$$

$$= \sum_{\substack{i_1, i_2, \\ j_1, j_2, \\ l, k}} x^{i_1 k} a^{k i_2} \delta^{j_1}{}_{i_1} \delta^{l}{}_{i_2} b^{l j_2} \frac{\partial}{\partial x^{j_1 j_2}} - (a \leftrightarrow b)$$

$$= \sum_{\substack{i_1, i_2, \\ l, k}} x^{i_1 k} [a^{kl} b^{l i_2} - b^{kl} a^{l i_2}] \frac{\partial}{\partial x^{i_1 i_2}}. \tag{8.35}$$

We see that $[A, B]$ is again an invariant vector field. The correspondence $A \mapsto (a^{i_1 i_2})$ mapping brackets of invariant vector fields onto commutators of matrices is a Lie algebra isomorphism.

8.4 From the representations of a Lie group to representations of its Lie algebra

A representation of a Lie group G on a manifold M is a group homeomorphism from G to Diff(M). A representation of the Lie algebra g on M is by definition a Lie algebra homomorphism from g, the infinitesimal version of G, to vect(M), the infinitesimal version of Diff(M).

To any representation ρ of G on M,

$$g \mapsto \rho_g \in \text{Diff}(M),$$

there corresponds a representation $\tilde{\rho}$ of g on M,

$$A \mapsto \tilde{\rho}_A \in \text{vect}(M),$$

given by

$$\tilde{\rho}_A(x) := T_e \text{ bit}_x A(e). \tag{8.36}$$

Note that the image $\tilde{\rho}(\mathfrak{g})$ is a finite-dimensional subalgebra of vect(M). The group representation is faithful if ρ is injective. Then also $\tilde{\rho}$ is injective. If ρ is free, all vector fields $\tilde{\rho}_A$ with $A \neq 0$ are nowhere vanishing.

In our example where $SO(3)$ is represented on S^2, it is natural to associate a vector field on the sphere with an infinitesimal rotation. This representation is not free, the vector field generated by rotation, say around the x^3-axis, vanishes at the north and south pole.

Let us consider the case of a linear representation ρ of G on a vector space V:

$$g \mapsto \rho_g \in GL(V) \subset \text{Diff}(V).$$

We define a linear vector field l on a vector space V to be a vector field whose components l^i, $i = 1, 2, \ldots, n$, with respect to the Cartesian coordinates y^i associated with a basis b_i of V, are linear functions:

$$l^i(v) = \sum_j l^i{}_j y^j, \quad v \in V, \quad l^i{}_j \in \mathbb{R}, \tag{8.37}$$

$$l = \sum_{i,j} l^i{}_j y^j \frac{\partial}{\partial y^i}. \tag{8.38}$$

The linear vector fields form a Lie subalgebra of $\mathrm{vect}\,(V)$ isomorphic to $gl(V)$. (The Lie algebra $gl(V)$ consists of all linear maps from V into itself with the commutator as the bracket operation.) Indeed

$$[l, l'] = \sum_{\substack{i,j, \\ i',j'}} l^i{}_j y^j \frac{\partial}{\partial y^i} l'^{i'}{}_{j'} y^{j'} \frac{\partial}{\partial y^{i'}} - (l \leftrightarrow l')$$

$$= \sum_{i,j,k} [l^i{}_k l'^k{}_j - l'^i{}_k l^k{}_j] y^j \frac{\partial}{\partial y^i}, \tag{8.39}$$

and with respect to the basis b_i (cf. (1.37)) a Lie algebra isomorphism is given by

$$l \mapsto -(l^i{}_j).$$

The minus sign is motivated by equation (8.41) below. If ρ is a linear representation of G, then all diffeomorphisms ρ_g are linear and it follows that all vector fields $\tilde{\rho}_A$ are linear:

$$A \mapsto \tilde{\rho}_A \in gl(V) \subset \mathrm{vect}\,(V).$$

To be specific, consider the fundamental representation of GL_n on $\mathbb{R}^n$. Let $x^{i_1 i_2}$ be the standard coordinates of GL_n and y^i the Cartesian coordinates of $\mathbb{R}^n$. In these coordinates the mapping

$$\mathrm{bit}_v : GL_n \to \mathbb{R}^n$$

reads

$$x^{i_1 i_2} \mapsto \sum_j (x^{-1})^{ij} y^j$$

with $v = (y^1, y^2, \ldots, y^n)$. If A is an element of gl_n given by

$$A(e) = \sum_{i_1, i_2} a^{i_1 i_2} \frac{\partial}{\partial x^{i_1 i_2}} \bigg|_e, \tag{8.40}$$

we calculate

$$\tilde{\rho}_A(v) = T_e \,\mathrm{bit}_v\, A(e)$$

$$= \sum_{\substack{i_1, i_2, \\ i,j}} a^{i_1 i_2} \frac{\partial}{\partial x^{i_1 i_2}} ((x^{-1})^{ij} y^j)|_{x^{ij} = \delta^i{}_j} \frac{\partial}{\partial y^i}$$

$$= -\sum_{i,j} a^{ij} y^j \frac{\partial}{\partial y^i}, \tag{8.41}$$

which is indeed a linear vector field.

An important example of a linear representation of a Lie group G is the adjoint representation. It acts on the vector space $V = T_e G$ (which is identified with the Lie algebra g) by:

$$\text{Ad}: G \rightarrow GL(T_e G)$$

$$g \mapsto \text{Ad}_g$$

$$\text{Ad}_g: T_e G \rightarrow T_e G$$

$$A(e) \mapsto T_e \text{aut}_g(A(e)).$$

It is well defined because for any g in G

$$\text{aut}_g(e) = geg^{-1} = e. \tag{8.42}$$

The corresponding linear representation of g on itself is also called adjoint representation and denoted by

$$\text{ad}_A := \widetilde{\text{Ad}}_A. \tag{8.43}$$

It follows that

$$\text{ad}_A B = [A, B], \tag{8.44}$$

where $[A, B]$ is the Lie bracket of invariant vector fields on G.

We end this section with two side remarks. Given any representation ρ of G on M, we can define an infinite-dimensional linear representation $\tilde{\rho}$ of G on vect(M) by

$$(\tilde{\rho}_g v)(x) := (T \rho_g)(v(\rho_g^{-1}(x))), \quad g \in G, x \in M, v \in \text{vect}(M), \tag{8.45}$$

which is well defined because ρ_g is a diffeomorphism. The subalgebra $\tilde{\rho}(g) \subset \text{vect}(M)$ is an invariant linear subspace,

$$\tilde{\rho}_g(\tilde{\rho}_A) = \tilde{\rho}_{\text{Ad}_g A}, \tag{8.46}$$

and therefore carries a finite-dimensional linear representation of G.

Similarly, we can always define a linear representation of the isotropy group at x on $V = T_x M$ by

$$I(x) \rightarrow GL(T_x M)$$

$$g \mapsto T_x \rho_g.$$

The adjoint representation is such a construction. It starts from the representation by inner automorphisms and uses the fact that the corresponding isotropy group at e is the entire group G. Also, it is in this sense that some physicists say: In the nonlinear realization of $SO(3)$ on S^2, $SO(2)$ is represented linearly.

8.5 From a Lie algebra to its Lie groups

THEOREM

Let G be a Lie group, $\mathfrak{g}$ its Lie algebra. There is a one-to-one corre-spondence between connected Lie subgroups of G and Lie subalgebras of $\mathfrak{g}$.

As a corollary we have: For every $A \in \mathfrak{g}$ there is a one-parameter subgroup

$$g: \mathbb{R} \to G$$

$$\tau \mapsto g_\tau,$$

$$g_0 = e, \tag{8.47}$$

$$g_{\tau+\sigma} = g_\tau g_\sigma \tag{8.48}$$

such that g is a curve with tangent vector $A(g_\tau)$ at g_τ.

We define the exponential map:

$$\exp: \mathfrak{g} \to G$$

$$A \mapsto \exp A := g_{\tau=1}. \tag{8.49}$$

Note the resemblance to the definition of geodesic coordinates. The exponential map is always differentiable but in general neither injective nor surjective. However locally, i.e. in some open neighbourhood of the origin in $\mathfrak{g}$, the exponential map is a diffeomorphism.

Two immediate properties are: First,

$$g_\tau = \exp \tau A. \tag{8.50}$$

Secondly, if $\tilde{\rho}_A \in \text{vect}(M)$ represents an element A of $\mathfrak{g}$ and $h_\tau \in \text{Diff}(M)$ is the flow of $\tilde{\rho}_A$, then

$$h_\tau = \rho_{\exp(-\tau A)}. \tag{8.51}$$

For the general linear group GL_n and its subgroups we have a convenient formula also justifying the name exponential map:

$$\exp A = \sum_{m=0}^{\infty} \frac{1}{m!} A^m, \quad A \in gl_n. \tag{8.52}$$

We have already used it in this form in equation (4.95).

Given a Lie algebra $\mathfrak{g}$, there is an infinite number of nonisomorphic Lie groups G with $\text{Lie}(G) = \mathfrak{g}$.

THEOREM

Let $\mathfrak{g}$ be a (finite-dimensional) Lie algebra. There is one and only one simply connected Lie group G with $\text{Lie}(G) = \mathfrak{g}$. For every other connected Lie group H with $\text{Lie}(H) = \mathfrak{g}$ there is a group homomorphism $\varphi: G \to H$ whose kernel is a discrete subgroup of G.

For a connected Lie group H, the simply connected Lie group G with $\text{Lie}(G) = \text{Lie}(H)$ is called the universal covering group of H.

As a first example consider $U(1)$. Its universal covering group is $\mathbb{R}$ and the group homomorphism $\varphi: \mathbb{R} \to U(1)$ is $\varphi(g) = \exp 2\pi i g$ whose kernel is the integers.

Another example is $SU(2)$, the universal covering group of $SO(3)$. Because of its importance in physics we shall consider it in detail. The groups $SO(3)$ and $SU(2)$ have isomorphic Lie algebras; a basis of $so(3)$ is:

$$T_1 = \begin{pmatrix} 0 & 0 & 0 \\ 0 & 0 & -1 \\ 0 & 1 & 0 \end{pmatrix}, \quad T_2 = \begin{pmatrix} 0 & 0 & 1 \\ 0 & 0 & 0 \\ -1 & 0 & 0 \end{pmatrix}, \quad T_3 = \begin{pmatrix} 0 & -1 & 0 \\ 1 & 0 & 0 \\ 0 & 0 & 0 \end{pmatrix}, \tag{8.53}$$

representing infinitesimal rotations around the x^1-, x^2-, and x^3-axis. The structure constants in this basis are read off from

$$[T_a, T_b] = \sum_c \varepsilon_{abc} T_c. \tag{8.54}$$

The real Lie algebra $su(2)$ consists of complex 2×2 matrices which are traceless and antihermitian. A convenient basis is $\tau_a = -\frac{1}{2}i\sigma_a$, $a = 1, 2, 3$, where σ_a are the Pauli matrices:

$$\sigma_1 = \begin{pmatrix} 0 & 1 \\ 1 & 0 \end{pmatrix}, \quad \sigma_2 = \begin{pmatrix} 0 & -i \\ i & 0 \end{pmatrix}, \quad \sigma_3 = \begin{pmatrix} 1 & 0 \\ 0 & -1 \end{pmatrix}. \tag{8.55}$$

The commutation relations are

$$[\tau_a, \tau_b] = \sum_c \varepsilon_{abc} \tau_c; \tag{8.56}$$

therefore $so(3)$ and $su(2)$ are isomorphic. Furthermore, the manifold of

$SU(2)$ is S^3, which is simply connected. The group $SO(3)$ is connected, but not simply connected. (Its manifold is the three-dimensional real projective space.)

Let us construct the homomorphism of the last theorem:

$$\varphi: SU(2) \rightarrow SO(3).$$

We need an auxiliary function

$$f: \mathbb{R}^3 \rightarrow su(2)$$

$$v = \begin{pmatrix} x^1 \\ x^2 \\ x^3 \end{pmatrix} \mapsto \sum_{j=1}^{3} x^j \tau_j = -\tfrac{1}{2}i \begin{pmatrix} x^3 & x^1 - ix^2 \\ x^1 + ix^2 & -x^3 \end{pmatrix}. \tag{8.57}$$

Let g be an $SU(2)$-matrix. Then we define $\varphi(g)$ by

$$\varphi(g)v := f^{-1}(g f(v) g^{-1}). \tag{8.58}$$

For all v the matrix $g f(v) g^{-1}$ is in $su(2)$ and $\varphi(g)$ is a well-defined linear map from $\mathbb{R}^3$ into itself. To show that it is indeed a rotation, we use

$$\det f(v) = \tfrac{1}{4}|v|^2 \tag{8.59}$$

and verify that $\varphi(g)$ preserves the length:

$$|\varphi(g)v|^2 = 4 \det (g f(v) g^{-1}) = 4 \det f(v) = |v|^2. \tag{8.60}$$

Furthermore, $\varphi(g)$ is orientation-preserving and therefore an element of $SO(3)$. The map φ is a surjective group homomorphism. However, it is not injective. To calculate its kernel, we look for all $g \in SU(2)$ such that $\varphi(g)v = v$ for all vectors v. This is equivalent to

$$g f(v) g^{-1} = f(v) \tag{8.61}$$

or

$$g\tau = \tau g \tag{8.62}$$

for all $\tau \in su(2)$. It follows that g must be a multiple of the identity matrix and $\ker \varphi = \{e, -e\}$. Consequently, φ is two-to-one:

$$\varphi(g) = \varphi(-g). \tag{8.63}$$

Note that in quantum mechanics, starting from the commutation relations of the angular momentum operators, we only know the Lie algebra $so(3) \cong su(2)$. Experiments tell us that not only the rotation group $SO(3)$ but also its universal covering group $SU(2)$ are represented in nature: Some interference experiments with neutrons, for instance in Grenoble (Rauch *et al.* 1975), do distinguish between g and $-g$ in $SU(2)$. The double

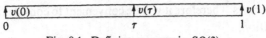

Fig. 8.1. Defining a curve in $SO(3)$.

valuedness of φ^{-1} stems from the fact that the rotation group is not simply connected. This can be visualized as follows. Take a long flexible strip, parametrize the long side by τ taking values between 0 and 1, call $v(\tau)$ the vector on the strip at τ pointing in the 'positive short' direction as in fig. 8.1. If we embed the strip arbitrarily in our $\mathbb{R}^3$, we can define a continuous map

$$Q:[0,1] \to SO(3)$$

$$\tau \mapsto Q(\tau),$$

where $Q(\tau)$ is the rotation that rotates $v(0)$ into $v(\tau)$ through their common plane. The family of rotations Q represents a closed curve in $SO(3)$ if we arrange the vectors $v(0)$ and $v(1)$ to be parallel. The curve is constant if the strip is flat; then all $v(\tau)$ are parallel. Now we twist the flat strip by keeping $v(0)$ fixed and turning $v(1)$ by 360°. Then it represents again a closed curve. Experimentally we can convince ourselves that there is no deformation of this strip keeping $v(0)$ and $v(1)$ parallel that transforms it into the flat strip. This closed curve cannot be shrunk to a point, the constant curve. On the other hand, there is such a deformation if we start from a situation prepared by twisting the flat strip by 720°. Locally φ^{-1} is a well-defined map from $SO(3)$ onto $SU(2)$ which we can use to lift our curves to $SU(2)$. Starting from $\varphi^{-1}(Q(0)) = \varphi^{-1}(\mathbb{1}) = +\mathbb{1} \in SU(2)$ we arrive, in the situation where v was twisted by 360°, at $\varphi^{-1}(Q(1)) = -\mathbb{1}$; i.e. $\varphi^{-1}(Q(\tau))$ is not closed. On the other hand, rotating v by 720° yields a closed curve also in $SU(2)$, which can be deformed into the constant curve because $SU(2)$ is simply connected.

We end this section with a remark on the representations of the universal covering group G of a connected Lie group H. Any representation ρ of H on M induces a representation $\rho \circ \varphi$ of G on M:

$$G \xrightarrow{\varphi} H \xrightarrow{\rho} \text{Diff}(M).$$

But G can have representations which do not come from a representation of H via φ, e.g. the fundamental representation of $SU(2)$. Other examples will be the subject of chapter 11.

8.6 The Maurer–Cartan form

We have defined the Lie algebra of a Lie group in terms of invariant vector fields. In the preceding chapters we have preferred to work with forms rather than vectors. Therefore it will be convenient to consider also here the dual formulation.

A differential form φ on a Lie group G is called (left) invariant if it is invariant under the pull pack by all left translations:

$$L_g^* \varphi = \varphi \tag{8.64}$$

for all $g \in G$. A left invariant form φ is uniquely determined by φ_e, an alternating form on $T_e G$:

$$\varphi_g = L_{g^{-1}}^* \varphi_e. \tag{8.65}$$

Let $A_1, \ldots, A_d$ be $d = \dim G$ linearly independent invariant vector fields on G. They form a basis of the Lie algebra $\mathfrak{g}$ and also a global frame on G. Let us denote by $\zeta^1, \ldots, \zeta^d$, $\zeta^i \in \Lambda^1 G$, the dual frame defined by

$$\zeta^i(A_j) = \delta^i{}_j. \tag{8.66}$$

It is a basis of the finite-dimensional vector space of left-invariant 1-forms. The ζ^i are called Maurer–Cartan forms. As the exterior derivative commutes with any pull back (cf. (7.57)), the 2-forms $d\zeta^i$ are also invariant and can be expanded

$$d\zeta^i = : -\tfrac{1}{2} \sum_{k,l} f_{kl}{}^i \zeta^k \wedge \zeta^l. \tag{8.67}$$

We now show that the numbers $f_{kl}{}^i$ are just the structure constants of $\mathfrak{g}$:

$$[A_k, A_l] = \sum_i f_{kl}{}^i A_i. \tag{8.68}$$

Indeed,

$$\begin{aligned} d\zeta^i(A_k, A_l) &= i_{A_l} i_{A_k} d\zeta^i \\ &= i_{A_l}(L_{A_k}\zeta^i - d i_{A_k}\zeta^i) = L_{A_k} i_{A_l}\zeta^i - i_{[A_k, A_l]}\zeta^i \\ &= -\sum_j f_{kl}{}^j \zeta^i(A_j) = -f_{kl}{}^i. \end{aligned} \tag{8.69}$$

We have used successively equations (1.23), (6.34), (8.66), and (6.35).

The identity

$$d^2 \zeta^i = 0 \tag{8.70}$$

is equivalent to the Jacobi identity expressed in terms of the structure constants. Consequently, the Lie algebra $\mathfrak{g}$ is uniquely characterized by the Maurer–Cartan forms ζ^i together with equation (8.67), the so-called Maurer–Cartan structure equation.

It is useful to repeat this construction in a basis independent manner. We define the Maurer–Cartan form ζ, a 1-form on G with values in $\mathfrak{g}$,

$$\zeta \in \Lambda^1(G, \mathfrak{g}),$$

by

$$\zeta|_{\mathfrak{g}} = \mathrm{id}_{\mathfrak{g}}, \tag{8.71}$$

i.e. ζ applied to an invariant vector field reproduces this invariant vector field. Since pointwise any vector field can be expanded in a frame of invariant vector fields, ζ is a well-defined 1-form and as such it is of course invariant under left translations. Under right translations, ζ transforms as

$$R_g^*\zeta = \mathrm{Ad}_{g^{-1}}\zeta. \tag{8.72}$$

The Maurer–Cartan structure equation (8.67) now reads

$$d\zeta = -\tfrac{1}{2}[\zeta, \zeta]. \tag{8.73}$$

Decomposing ζ with respect to the basis A_i of $\mathfrak{g}$ we retrieve real-valued forms ζ^i:

$$\zeta = \sum_i \zeta^i A_i. \tag{8.74}$$

As an example we compute the Maurer–Cartan form for GL_n in the standard coordinates $x^{i_1 i_2}$. In these coordinates an invariant vector field $A \in \mathrm{Lie}(GL_n)$ can be written:

$$A(g) = \sum_{\substack{j_1, j_2, \\ l}} x^{j_1 l}(g) a^{l j_2} \frac{\partial}{\partial x^{j_1 j_2}}. \tag{8.75}$$

Using the identification of $A(g)$ with the gl_n-matrix $(a^{i_1 i_2})$, $\zeta(e)$ becomes

$$\zeta^{i_1 i_2}(e) = dx^{i_1 i_2}|_e. \tag{8.76}$$

Indeed

$$\zeta^{i_1 i_2}|_e(A(e)) = \sum_{j_1, j_2} a^{j_1 j_2} dx^{i_1 i_2}|_e\left(\frac{\partial}{\partial x^{j_1 j_2}}\bigg|_e\right) = a^{i_1 i_2} \tag{8.77}$$

and

$$\zeta^{i_1 i_2}(g) = L_{g^{-1}}^* \zeta^{i_1 i_2}(e) = x^{i_1 l}(g^{-1}) dx^{l i_2}|_g. \tag{8.78}$$

Like the shorthand $A(g) = gA(e)$, this formula is often abbreviated as

$$\zeta(g) = g^{-1} dg \tag{8.79}$$

and applies also to subgroups of GL_n. Note the resemblance with the inhomogeneous term in the transformation law of a connection, equation (4.42). However, here the exterior derivative refers to GL_n and not to spacetime.

Problems

8.1 Show that $\mathbb{R}^3$ with the multiplication law

$$\begin{pmatrix} x_1 \\ x_2 \\ x_3 \end{pmatrix} \begin{pmatrix} y_1 \\ y_2 \\ y_3 \end{pmatrix} = \begin{pmatrix} x_1 + y_1 \\ x_2 + y_2 \\ x_3 + y_3 + x_1 y_2 - x_2 y_1 \end{pmatrix}$$

is a Lie group and that its left action is an affine representation. Calculate the structure constants of the Lie algebra with respect to a convenient basis.

8.2 Show that the Lie algebra of $SO(n)$ is isomorphic to the Lie algebra of antisymmetric matrices.

8.3 Define $C \in gl_n$ by

$$e^C = e^{\lambda A} e^{\lambda B}$$

with $A, B \in gl_n$, and $\lambda \in \mathbb{R}$. Consider the expansion of C in powers of λ and calculate as many terms as you wish (Campbell–Hausdorff formula).

8.4 Use the Maurer–Cartan structure equation (8.67) to prove that (8.70) is equivalent to the Jacobi identity.

9

Fibre bundles

We have already used the word 'bundle' occasionally. Think, for example, of the tangent bundle of an open subset $\mathcal{U}$ of $\mathbb{R}^n$: A vector space isomorphic to $\mathbb{R}^n$, namely the tangent space $T_x\mathcal{U}$, is attached to each point $x \in \mathcal{U}$. In this case the bundle is simply $\mathcal{U} \times \mathbb{R}^n$, the Cartesian product of the 'base space' $\mathcal{U}$ and the 'fibre' $\mathbb{R}^n$. An obvious generalization is to replace $\mathcal{U}$ by an arbitrary differentiable manifold M and similarly the fibre $\mathbb{R}^n$ by some other manifold F. What is less obvious, however, is how to twist the fibres such that the resulting space is not the Cartesian product $M \times F$. The tools for dealing with such spaces are provided in this chapter. Furthermore, we shall see that many of the mathematical and physical concepts introduced so far extend naturally to this more complicated situation.

9.1 The notion of fibre bundles

In order to describe a fibre bundle we need manifolds B (the bundle space or total space), M (the base space), F (the standard fibre), and a Lie group G (the structure group) which is represented effectively on F. So for each $g \in G$ we have a diffeomorphism

$$\rho_g : F \to F \tag{9.1}$$

and one can identify g with ρ_g. Furthermore, we need a surjective (and, of course, differentiable) map

$$\pi : B \to M, \tag{9.2}$$

called the bundle projection, such that $\pi^{-1}(x)$, the fibre over x, is diffeomorphic to F for all $x \in M$ (see fig. 9.1).

In addition, we demand that the bundle be locally trivial. This means: There is an open covering $\{\mathcal{U}_r\}$ of M such that $\pi^{-1}(\mathcal{U}_r)$ is diffeomorphic to $\mathcal{U}_r \times F$ via a diffeomorphism of the form

$$\pi^{-1}(\mathcal{U}_r) \to \mathcal{U}_r \times F$$

$$p \mapsto (\pi(p), f_r(p)) \tag{9.3}$$

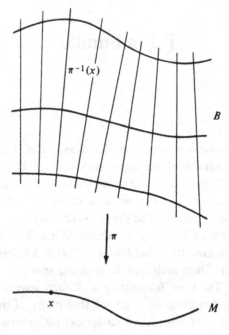

Fig. 9.1. Fibre bundle.

where

$$f_r: \pi^{-1}(\mathcal{U}_r) \to F. \tag{9.4}$$

Therefore, $\pi^{-1}(\mathcal{U}_r)$, the portion of the bundle over $\mathcal{U}_r$, can be identified with the Cartesian product $\mathcal{U}_r \times F$. The diffeomorphism (9.3) is called a local trivialization and $\mathcal{U}_r$ a trivializing neighbourhood.

Restricting the map f_r to the fibre over $x \in \mathcal{U}_r$, we arrive at a diffeomorphism

$$f_{r,x} := f_r|_{\pi^{-1}(x)}: \pi^{-1}(x) \to F \tag{9.5}$$

identifying $\pi^{-1}(x)$ with the standard fibre. Note that this identification is not canonical. For $x \in \mathcal{U}_r \cap \mathcal{U}_s$ we have two different identifications related by the diffeomorphism

$$f_{s,x} \circ f_{r,x}^{-1}: F \to F \tag{9.6}$$

which is required to correspond to an element $g_{sr}(x)$ of the structure group G via the representation ρ. So we get the transition functions

$$\begin{aligned} g_{sr}: \mathcal{U}_r \cap \mathcal{U}_s &\to G \\ x &\mapsto g_{sr}(x) \end{aligned} \tag{9.7}$$

where

$$\rho(g_{sr}(x)) = f_{s,x} \circ f_{r,x}^{-1}, \tag{9.8}$$

or phrased differently,

$$\rho(g_{sr}(x))f_{r,x}(p) = f_{s,x}(p)$$

for $p \in \pi^{-1}(x)$. The maps g_{sr} are required to be smooth and indicate how the spaces $\mathcal{U}_r \times F$ and $\mathcal{U}_s \times F$, each of which can be identified with a part of the bundle, are to be glued together. They satisfy the so-called cocycle condition

$$g_{sr}(x)g_{rt}(x) = g_{st}(x) \tag{9.9}$$

for $x \in \mathcal{U}_r \cap \mathcal{U}_s \cap \mathcal{U}_t$. In particular, we have

$$g_{sr}(x) = g_{rs}(x)^{-1}. \tag{9.10}$$

Conversely, if F and ρ are given, then for any family of G-valued functions g_{sr} such that (9.9) holds there is a unique fibre bundle whose transition functions are g_{sr}.

We shall denote the bundle just described by (B, M, π). But often we shall call the total space B alone a bundle or a G-bundle over M, if we want to be more specific.

The bundle (B, M, π) is called trivial if M itself is a trivializing neighbourhood. It may then be identified with $M \times F$. One can prove that any bundle over a contractible base space like, for example, $\mathbb{R}^n$ is trivial.

Let us mention two important special cases:

(a) Vector bundles. Here the fibres are vector spaces, the structure group is a subgroup of $GL(F)$, and the diffeomorphisms $f_{r,x}$ are vector space isomorphisms.

(b) Principal bundles. In this case, $F = G$ with the further requirement that G is represented on itself by left translation: $\rho_g = L_g$.

9.2 The right action of the structure group on a principal bundle

On a principal bundle we have a natural right action of the structure group G, denoted by

$$\tilde{R}_g : P \to P, \quad g \in G. \tag{9.11}$$

It is defined as follows. Let $\{\mathcal{U}_r\}$ be an open covering of M consisting of trivializing neighbourhoods as above. For $p \in \pi^{-1}(x)$ with $x \in \mathscr{U}_r$ we put $g_r = f_{r,x}(p) \in G$ and define

$$(\tilde{R}_g p)_r := f_{r,x}^{-1}(R_g g_r) = f_{r,x}^{-1}(g_r g); \qquad (9.12)$$

i.e., under a local trivialization, identifying the fibre over x with G, the right action $\tilde{R}$ corresponds to right translation.

We have added a subscript r to $\tilde{R}_g p$ because we used a trivialization of the bundle over $\mathscr{U}_r$ in our definition. It is, however, easy to show that $(\tilde{R}_g p)_r$ is actually independent of the particular trivialization. To see this take $x \in \mathscr{U}_r \cap \mathscr{U}_s$ and $p \in \pi^{-1}(x)$. Under the local trivializations over $\mathscr{U}_r$ and $\mathscr{U}_s$ the point p corresponds to g_r and g_s, respectively:

$$g_r = f_{r,x}(p), \quad g_s = f_{s,x}(p).$$

The group elements g_r and g_s representing the same $p \in P$ are related by the transition function g_{rs}:

$$g_r = f_{r,x}(f_{s,x}^{-1}(g_s)) = g_{rs}(x)g_s. \qquad (9.13)$$

So we get the desired result:

$$(\tilde{R}_g p)_s = f_{s,x}^{-1}(g_s g) = f_{r,x}^{-1} \circ f_{r,x} \circ f_{s,x}^{-1}(g_s g)$$
$$= f_{r,x}^{-1}(g_{rs}(x)g_s g) = f_{r,x}^{-1}(g_r g) = (\tilde{R}_g p)_r. \qquad (9.14)$$

Therefore we are allowed to omit the index r of $\tilde{R}_g p$. Sometimes we shall also write pg instead of $\tilde{R}_g p$. Note how the independence of $\tilde{R}_g p$ of the particular trivialization in (9.12) is connected to the fact that the structure group G is represented on the standard fibre ($= G$) by left translation. Obviously, $\tilde{R}_g$ maps each fibre onto itself,

$$\pi(\tilde{R}_g p) = \pi(p), \qquad (9.15)$$

and defines a free, transitive action of G on $\pi^{-1}(x)$ for all $x \in M$.

9.3 Reduction of the structure group

In our definition of a fibre bundle the structure group G appeared only as the space in which the transition functions take their values. So one could often replace G by a larger group $\tilde{G} \supset G$. The opposite procedure is less trivial. It may be possible for a given bundle with structure group G to find a set of local trivializations such that the corresponding transition functions are G'-valued, where G' is a Lie subgroup of G. If this is the case, one says that the structure group G is reducible to G'. For example, the structure group of a trivial bundle is reducible to the trivial group $\{e\}$.

9.4 Bundle maps

We consider only the two special cases of principal bundles and vector bundles.

Let (P, M, π) and (P', M', π') be principal bundles with structure groups G and G', respectively. A bundle map is a triple (f_P, φ, f_M) of mappings

$$f_P: P \to P', \quad \varphi: G \to G', \quad f_M: M \to M'$$

such that

(a) φ is a group homomorphism;
(b) $f_M \circ \pi = \pi' \circ f_P$;
(c) $f_P \circ \tilde{R}_g = \tilde{R}_{\varphi(g)} \circ f_P$ for all $g \in G$.

Condition (b) can also be expressed as the commutativity of the diagram

$$
\begin{array}{ccc}
P & \xrightarrow{\;f_P\;} & P' \\
\pi \downarrow & & \downarrow \pi' \\
M & \xrightarrow{\;f_M\;} & M'
\end{array}
\qquad (9.16)
$$

and means that each fibre of P is mapped into a fibre of P'. Condition (c) demands the compatibility of f_P with the right actions of the structure groups.

For two vector bundles (E, M, π) and (E', M', π') we define a bundle map to be a pair (f_E, f_M) of mappings

$$f_E: E \to E', \quad f_M: M \to M'$$

such that fibres are mapped into fibres; i.e. $f_M \circ \pi = \pi' \circ f_E$, and f_E restricted to any fibre is a linear map of vector spaces. Again, we have a commuting diagram like (9.16):

If the mappings defining a bundle map are diffeomorphisms, it is called a bundle isomorphism.

9.5 Examples

Our first example of a fibre bundle is a trivial bundle where the total space B is the Cartesian product of the base space M and the standard fibre F. The projection

$$\pi: M \times F \to M \qquad (9.17)$$

is simply the projection onto the first component so that $\pi^{-1}(x) = \{x\} \times F$.

One trivialization with $\mathcal{U} = M$ and

$$f : \pi^{-1}(M) = M \times F \to F$$
$$(x, b) \mapsto b$$

(9.18)

is already sufficient. Covering M by several trivializing neighbourhoods $\mathcal{U}_r$ and taking $f_r := f|_{\pi^{-1}(\mathcal{U}_r)}$, one sees that the structure group is indeed reducible to the trivial group. Choosing f_rs which are not simply restrictions of one global f, we could, however, arrive at an arbitrarily complicated description of a trivial bundle simulating a larger structure group.

As a second example we discuss the tangent bundle TM of an n-dimensional manifold M. It is a vector bundle with bundle space $TM = \cup_x T_x M$, base space M, and standard fibre $\mathbb{R}^n$. The structure group is GL_n (or a subgroup of GL_n) acting on the standard fibre by matrix multiplication:

$$\rho_g : \mathbb{R}^n \to \mathbb{R}^n$$
$$v \mapsto gv$$

(9.19)

for $g \in GL_n$. The bundle projection π maps a tangent vector at $x \in M$ onto x. Consequently, the fibre over x is $\pi^{-1}(x) = T_x M$.

Local trivializations can be constructed with the help of an atlas $\{(\mathcal{U}_r, \alpha_r)\}$ for M. Denoting the coordinates on $\mathcal{U}_r$ by $x_r^1, x_r^2, \ldots, x_r^n$, we write a tangent vector $v \in \pi^{-1}(\mathcal{U}_r)$ with $\pi(v) = x$, i.e. $v \in T_x M$, as

$$v = \sum_{i=1}^n v_r^i \frac{\partial}{\partial x_r^i}\bigg|_x .$$

(9.20)

We define

$$f_r(v) := (v_r^1, v_r^2, \ldots, v_r^n) \in \mathbb{R}^n.$$

(9.21)

If $\pi(v) = x \in \mathcal{U}_r \cap \mathcal{U}_s$ one finds (see (7.35))

$$v_r^j = \sum_{i=1}^n \frac{\partial x_r^j}{\partial x_s^i} v_s^i$$

so that

$$f_{r,x}(v) = g_{rs}(x) f_{s,x}(v)$$

(9.22)

with the transition function $g_{rs}(x) \in GL_n$:

$$(g_{rs}(x))^j{}_i = \frac{\partial x_r^j}{\partial x_s^i}.$$

(9.23)

Taking as coordinates of $v \in \pi^{-1}(x) = T_x M \subset TM$ with $x \in \mathcal{U}_r$, the $2n$-tuple of real numbers $(x_r^1, x_r^2, \ldots, x_r^n, v_r^1, v_r^2, \ldots, v_r^n)$, we see from the above that the changes of coordinates are differentiable. In this way TM becomes a $2n$-dimensional differentiable manifold, and the local trivializations

$$\pi^{-1}(\mathcal{U}_r) \to \mathcal{U}_r \times \mathbb{R}^n$$
$$v \mapsto (\pi(v), f_r(v)) \tag{9.24}$$

are indeed diffeomorphisms.

Remember that M is orientable if and only if it possesses an atlas such that all changes of coordinates have positive Jacobian. Due to (9.23) we can phrase this statement in bundle language as: M is orientable if and only if the structure group of TM may be reduced to GL_n^+.

Our next example is a principal bundle, the frame bundle $F(M)$ of an n-dimensional manifold M. The bundle space is

$$F(M) = \{(x, b_1, b_2, \ldots, b_n) | x \in M; b_1, b_2, \ldots, b_n \text{ basis of } T_x M\},$$

the base space is M, and the projection π is defined by

$$\pi: F(M) \to M$$
$$(x, b_1, b_2, \ldots, b_n) \mapsto x. \tag{9.25}$$

The fibre over x, the set of all bases of $T_x M$, can be identified with GL_n: Let $\bar{b}_1, \bar{b}_2, \ldots, \bar{b}_n$ be a fixed basis of $T_x M$. Then an arbitrary basis $b_1, b_2, \ldots, b_n$ can be expanded as

$$b_i = \sum_{j=1}^n \gamma^j{}_i \bar{b}_j \tag{9.26}$$

determining a unique $\gamma \in GL_n$. Therefore the standard fibre is GL_n.

In order to see that GL_n is also the structure group and as such acts on the standard fibre by left translation, we construct local trivializations. Let $\{(\mathcal{U}_r, \alpha_r)\}$ be an atlas for M and denote the coordinates on $\mathcal{U}_r$ by $x_r^1, x_r^2, \ldots, x_r^n$. For $(x, b_1, b_2, \ldots, b_n) \in \pi^{-1}(\mathcal{U}_r)$ we define

$$f_r(x, b_1, b_2, \ldots, b_n) := (dx_r^j(b_i)) \in GL_n. \tag{9.27}$$

Since

$$b_i = \sum_{j=1}^n dx_r^j(b_i) \frac{\partial}{\partial x_r^j}\bigg|_x \tag{9.28}$$

this means that we choose as our reference basis $\bar{b}_j = \partial/\partial x_r^j|_x$. Using (2.29) it is easy to calculate the transition functions:

$$f_{r,x}(x, b_1, b_2, \ldots, b_n) = (dx_r^j(b_i)) = \left(\sum_{k=1}^n \frac{\partial x_r^j}{\partial x_s^k} dx_s^k(b_i) \right)$$

$$= \left(\sum_{k=1}^n (g_{rs}(x))^j{}_k dx_s^k(b_i) \right)$$

$$= g_{rs}(x) f_{s,x}(x, b_1, b_2, \ldots, b_n). \tag{9.29}$$

So they turn out to coincide with the transition functions (9.23) of TM: TM is a vector bundle associated to $F(M)$ (see section 9.6 below). Moreover, (9.29) shows us that the structure group is represented on the standard fibre by left translation as it should be for a principal bundle. A differentiable atlas for $F(M)$ is easily exhibited giving $F(M)$ the structure of an $(n + n^2)$-dimensional manifold.

Let us work out the right action of GL_n on $F(M)$. For $\gamma \in GL_n$, $x \in \mathcal{U}_r$ and $b_1, b_2, \ldots, b_n$ a basis of $T_x M$,

$$\tilde{R}_\gamma(x, b_1, b_2, \ldots, b_n) = (x, b_1, b_2, \ldots, b_n)\gamma$$

is defined such that (cf. section 9.2)

$$f_{r,x}((x, b_1, b_2, \ldots, b_n)\gamma) = f_{r,x}(x, b_1, b_2, \ldots, b_n)\gamma. \tag{9.30}$$

On the rhs of the last equation matrix multiplication is understood. So we get

$$(f_{r,x}((x, b_1, b_2, \ldots, b_n)\gamma))^j{}_i = \sum_{k=1}^n dx_r^j(b_k)\gamma^k{}_i$$

$$= (f_{r,x}(x, b_1', b_2', \ldots, b_n'))^j{}_i \tag{9.31}$$

with

$$b_i' = \sum_{k=1}^n \gamma^k{}_i b_k \tag{9.32}$$

and therefore

$$(x, b_1, b_2, \ldots, b_n)\gamma = \left(x, \sum_{i=1}^n \gamma^i{}_1 b_i, \sum_{i=1}^n \gamma^i{}_2 b_i, \ldots, \sum_{i=1}^n \gamma^i{}_n b_i \right). \tag{9.33}$$

In the case of the bundles considered so far the fibre structure was obvious while the bundle space was an abstract construction. In our next example we have the opposite situation. The total space is S^3. But to find a projection π which makes it a $U(1)$ principal bundle over S^2 requires some thought. Consider S^3 as a subset of $\mathbb{C}^2$:

$$S^3 = \{(z_1, z_2) \in \mathbb{C}^2 \,||z_1|^2 + |z_2|^2 = 1\}. \tag{9.34}$$

The so-called Hopf map $\pi: S^3 \to S^2$ is defined by

$$\pi(z_1, z_2) = (\bar{z}_1 z_2 + \bar{z}_2 z_1, -i\bar{z}_1 z_2 + i\bar{z}_2 z_1, |z_1|^2 - |z_2|^2). \qquad (9.35)$$

Parametrizing S^3 by

$$z_1 = \cos\tfrac{1}{2}\vartheta \ e^{i\psi_1}, \quad z_2 = \sin\tfrac{1}{2}\vartheta \ e^{i\psi_2} \qquad (9.36)$$

with $0 \leqslant \vartheta \leqslant \pi$ and $\psi_1, \psi_2 \in \mathbb{R}$, we get

$$\begin{aligned}
&\pi(\cos\tfrac{1}{2}\vartheta \ e^{i\psi_1}, \sin\tfrac{1}{2}\vartheta \ e^{i\psi_2}) \\
&= (\sin\vartheta\cos(\psi_2 - \psi_1), \sin\vartheta\sin(\psi_2 - \psi_1), \cos\vartheta).
\end{aligned} \qquad (9.37)$$

So π indeed maps S^3 onto S^2, and we see that $\pi(z_1, z_2) = \pi(z_1', z_2')$ if and only if $z_j' = e^{i\alpha} z_j$ $(j = 1, 2)$ with $\alpha \in \mathbb{R}$. Therefore the standard fibre is $U(1)$.

As trivializing neighbourhoods we can take

$$\mathcal{U}_1 = S^2 - \{(0, 0, -1)\}, \quad \mathcal{U}_2 = S^2 - \{(0, 0, 1)\}. \qquad (9.38)$$

On $\pi^{-1}(\mathcal{U}_r) = \{(z_1, z_2) \in S^3 | z_r \neq 0\}$ we define $f_r: \pi^{-1}(\mathcal{U}_r) \to U(1)$ by

$$f_r(z_1, z_2) := z_r / |z_r|. \qquad (9.39)$$

For $x = (\sin\vartheta\cos\varphi, \sin\vartheta\sin\varphi, \cos\vartheta) \in \mathcal{U}_1 \cap \mathcal{U}_2$ and $\alpha \in \mathbb{R}$ one finds

$$f_{1,x}^{-1}(e^{i\alpha}) = (\cos\tfrac{1}{2}\vartheta \ e^{i\alpha}, \sin\tfrac{1}{2}\vartheta \ e^{i(\varphi + \alpha)}) \qquad (9.40)$$

and consequently

$$(f_{2,x} \circ f_{1,x}^{-1})(e^{i\alpha}) = e^{i\varphi} e^{i\alpha}. \qquad (9.41)$$

Therefore the transition function relative to our trivialization is given by

$$\begin{aligned}
&g_{21}: \mathcal{U}_1 \cap \mathcal{U}_2 \to U(1) \\
&(\sin\vartheta\cos\varphi, \sin\vartheta\sin\varphi, \cos\vartheta) \mapsto e^{i\varphi}.
\end{aligned} \qquad (9.42)$$

The image of any parallel on S^2 under g_{21} winds once around the $U(1)$-circle.

Equation (9.41) shows that the structure group $U(1)$ acts on the standard fibre $U(1)$ by left translation, and we have indeed a $U(1)$ principal bundle over S^2. For the right action of $U(1)$ on S^3, the total space of our bundle, we find

$$\tilde{R}_{e^{i\alpha}}(z_1, z_2) = (z_1 e^{i\alpha}, z_2 e^{i\alpha}), \quad (z_1, z_2) \in S^3, \quad \alpha \in \mathbb{R}. \qquad (9.43)$$

The bundle just described will reappear in the discussion of the Dirac monopole (see chapter 10).

As a side remark we mention the following theorem. Let G be a Lie group and H a closed Lie subgroup. Then we can consider G as a principal bundle with base space G/H (left cosets) and structure group H such that the bundle projection $\pi: G \to G/H$ is the canonical projection. The right action of H on G is given by

$$\tilde{R}_h g = R_h g = gh, \quad g \in G, \quad h \in H.$$

In particular, $SU(n)$ is an $SU(n-1)$ bundle over S^{2n-1} ($n \geqslant 3$) and $U(n)$ is a $U(n-1)$ bundle over S^{2n-1} ($n \geqslant 2$).

9.6 Associated bundles

Let (B, M, π_B) be a fibre bundle with standard fibre F and (P, M, π) a principal bundle, both with the same base space M and the same structure group G. Then B is called associated to P, if there is an open covering $\{\mathcal{U}_r\}$ of M consisting of neighbourhoods which trivialize P as well as B such that the corresponding transition functions for P and B coincide.

The most important case is that of a vector bundle B: Here F is a vector space on which the G-valued transition functions act via some linear representation, whereas on the fibres of P they are represented by left translation. We have already seen that the tangent bundle TM is associated with the frame bundle $F(M)$.

9.7 Sections

A section of a fibre bundle (B, M, π) is a map $\sigma: M \to B$ such that $\pi \circ \sigma = \mathrm{id}_M$, i.e., $x \in M$ is mapped onto a point $\sigma(x)$ in the fibre over x. For example, a section of the tangent bundle of a manifold is a vector field.

Call the standard fibre of our bundle F. Under a local trivialization

$$\pi^{-1}(\mathcal{U}_r) \to \mathcal{U}_r \times F$$
$$p \mapsto (\pi(p), f_r(p)) \tag{9.44}$$

we get for $x \in \mathcal{U}_r$:

$$\sigma(x) \mapsto (\pi(\sigma(x)), f_r(\sigma(x))) = (x, (f_r \circ \sigma)(x)). \tag{9.45}$$

So, locally a section may be described by an F-valued function on the base space, here $f_r \circ \sigma$. These local functions are glued together by means of the transition functions. For $x \in \mathcal{U}_r \cap \mathcal{U}_s$ we have

$$(f_s \circ \sigma)(x) = f_{s,x}(\sigma(x)) = \rho(g_{sr}(x)) f_{r,x}(\sigma(x))$$
$$= \rho(g_{sr}(x))(f_r \circ \sigma)(x), \tag{9.46}$$

where ρ is the representation of G on F.

Local sections (defined only on subsets of M) exist always. Let, for example, $h: \mathcal{U}_r \to F$ be an arbitrary map. Then

$$\mathcal{U}_r \to \pi^{-1}(\mathcal{U}_r)$$

$$x \mapsto f_{r,x}^{-1}(h(x)) \tag{9.47}$$

is a section over $\mathcal{U}_r$. On the other hand, sections defined on all of M do not necessarily exist.

However, any vector bundle (E, M, π) admits sections. Let, for example, $\{\mathcal{U}_r\}$ be a locally finite trivializing covering of M, $\{h_r\}$ a partition of unity subordinate to that covering and $\{\sigma_r\}$ a family of local sections

$$\sigma_r: \mathcal{U}_r \to \pi^{-1}(\mathcal{U}_r). \tag{9.48}$$

Then

$$\sigma(x) := \sum_r h_r(x)\sigma_r(x) \tag{9.49}$$

defines a section of E. Note how the vector space structure of the fibres enters the construction. In particular, we have the zero section

$$x \mapsto 0 \in \pi^{-1}(x), \tag{9.50}$$

which is well defined because 0 is invariant under the action of the structure group.

For principal bundles the situation is different.

THEOREM

A principal bundle (P, M, π) is trivial if and only if it possesses a section.

Proof: Let us denote the structure group by G. If P is trivial, say $P = M \times G$, we can define a section by

$$\sigma(x) := (x, e) \tag{9.51}$$

for all $x \in M$. Conversely, assume that we have a section $\sigma: M \to P$. Since the right action of G on P is transitive and free on each fibre, every $p \in \pi^{-1}(x)$ can be written as $p = \sigma(x)g_0$ with a unique $g_0 \in G$. Consequently, we may trivialize P by mapping p onto $(x, g_0) \in M \times G$. QED

Since a global frame on a manifold is a section of the frame bundle, we get as a corollary another characterization of parallelizable manifolds: A manifold is parallelizable if and only if its frame bundle is trivial.

9.8 Connections on principal bundles

We consider a principal bundle (P, M, π) with structure group G. The fibration of P by means of the projection π allows us to introduce the concept of vertical tangent vectors on P: A tangent vector $v \in T_p P$ is called vertical if $T_p \pi(v) = 0$. By V_p we denote the subspace of vertical tangent vectors in $T_p P$. The right action $\tilde{R}$ of G on P leads to the maps

$$\text{bit}_p : G \to P$$
$$g \mapsto \tilde{R}_g p = pg \qquad (9.52)$$

for $p \in P$, and to a Lie algebra homomorphism

$$\lambda : \mathfrak{g} \to \text{vect}\,(P)$$
$$A \mapsto \lambda A \qquad (9.53)$$

defined by

$$(\lambda A)(p) := T_e \text{bit}_p(A(e)). \qquad (9.54)$$

(Comparing with the formulas of chapter 8 one has to take into account that ρ in (8.18), for instance, is a left action.) The vector fields λA on P, called fundamental vector fields, are vertical: $(\lambda A)(p) \in V_p$ for all $p \in P$. This is easily shown. Since for all $g \in G$

$$\pi \circ \text{bit}_p(g) = \pi(pg) = \pi(p), \qquad (9.55)$$

$\pi \circ \text{bit}_p$ is constant on G. Consequently,

$$T_p \pi((\lambda A)(p)) = T_p \pi \circ T_e \text{bit}_p(A(e)) = T_e(\pi \circ \text{bit}_p)(A(e)) = 0. \qquad (9.56)$$

Each map bit_p provides an identification of G and the fibre to which p belongs, and $T_e \text{bit}_p$ does the analogous job for $T_e G$ (or $\mathfrak{g}$) and V_p: The correspondence

$$A \leftrightarrow (\lambda A)(p) \qquad (9.57)$$

is a (vector space) isomorphism of $\mathfrak{g}$ and V_p. So we may consider vertical tangent vectors as being tangent to the fibre. In particular, $\dim V_p = \dim G$.

For a trivial bundle $P = M \times G$ (or locally in an arbitrary bundle) we find the following formulas. Write $p \in P$ as $p = (x, h) \in M \times G$. Then $\pi(x, h) = x$ and a tangent vector $v \in T_p P = T_{(x,h)} M \times G = T_x M \times T_h G$ can be decomposed as (see section 7.7)

$$v = (v_M, v_G) \in T_x M \times T_h G. \qquad (9.58)$$

Since $T_p \pi(v) = v_M$, we get

$$V_p = \{(0_M, v_G) | v_G \in T_h G\}, \qquad (9.59)$$

where 0_M denotes the zero vector in $T_x M$. With

$$i_x: G \to P$$
$$g \mapsto (x, g) \qquad (9.60)$$

we can write

$$\text{bit}_p(g) = (x, hg) = i_x(hg)$$

or

$$\text{bit}_p = i_x \circ L_h. \qquad (9.61)$$

So we obtain

$$(\lambda A)(p) = T_e \, \text{bit}_p(A(e)) = T_h i_x(T_e L_h(A(e)))$$
$$= T_h i_x(A(h)) = (0_M, A(h)). \qquad (9.62)$$

The definition of the vertical tangent spaces V_p is an immediate consequence of the characterizing properties of a fibre bundle. Complementary subspaces of $T_p P$, on the other hand, are not canonical but must be introduced as an additional structure, if desired. Requiring compatibility with the already existing structures, we are led to the following definition.

A connection on the principal bundle (P, M, π) is a family of subspaces $H_p \subset T_p P$, called horizontal tangent spaces, such that

(a) H_p is complementary to V_p in $T_p P$;
(b) H_p depends smoothly on p; i.e., H_p is locally spanned by smooth vector fields on P;
(c) $H_{pg} = T_p \tilde{R}_g(H_p)$ for all $g \in G$.

Once we have a connection, we can decompose every tangent vector $v \in T_p P$ uniquely into a horizontal and a vertical part:

$$v = Vv + Hv, \quad Vv \in V_p, \quad Hv \in H_p. \qquad (9.63)$$

Condition (c) implies that the right action of G on P commutes with this decomposition. For we have

$$T_p \tilde{R}_g(Hv) \in H_{pg} \qquad (9.64)$$

because of (c). Furthermore,

$$T_{pg}\pi(T_p \tilde{R}_g(Vv)) = T_p(\pi \circ \tilde{R}_g)(Vv) = T_p\pi(Vv) = 0 \qquad (9.65)$$

so that

$$T_p \tilde{R}_g(Vv) \in V_{pg}. \qquad (9.66)$$

Since trivially

$$T_p \tilde{R}_g(v) = T_p \tilde{R}_g(Vv) + T_p \tilde{R}_g(Hv) \qquad (9.67)$$

we conclude

$$VT_p \tilde{R}_g(v) = T_p \tilde{R}_g(Vv),$$
$$HT_p \tilde{R}_g(v) = T_p \tilde{R}_g(Hv). \qquad (9.68)$$

One can prove that a connection exists on any principal bundle. The geometrical meaning of a connection lies in the fact that it may be used to define parallel translation (see section 9.12).

It is convenient to replace the above description of a connection by a formulation employing differential forms. The 1-form $\mathscr{A}$ of a connection is the g-valued 1-form on P which for $v \in T_p P$ is given by

$$\mathscr{A}_p(v) = B \in \mathfrak{g}, \qquad (9.69)$$

where B is such that $(\lambda B)(p) = Vv$. So $\mathscr{A}_p(v)$ is the element of g which under the isomorphism (9.57) corresponds to the vertical part of v, and we can write

$$Vv = (\lambda \mathscr{A}_p(v))(p). \qquad (9.70)$$

Restricting our attention to one fibre $\pi^{-1}(x)$ we may identify it with G by means of bit_{p_0}, where $p_0 \in \pi^{-1}(x)$ is fixed. Then only vertical tangent vectors appear, which can be written as $(\lambda B)(p)$ with $B \in \mathfrak{g}$ and we get

$$\mathscr{A}_{p_0 g}((\lambda B)(p_0 g)) = B. \qquad (9.71)$$

Apart from the homomorphism λ identifying left-invariant vector fields on G (Lie algebra elements) and fundamental vector fields on P, this is precisely the definition (8.71) of the Maurer–Cartan form.

Let us collect some properties of $\mathscr{A}$:

(a) $\mathscr{A}_p(H_p) = 0;$ (9.72)

(b) $\mathscr{A}_p((\lambda B)(p)) = B$ for all $p \in P,\ B \in \mathfrak{g};$ (9.73)

(c) $(\tilde{R}_g^* \mathscr{A})_p(v) = \mathrm{Ad}(g^{-1}) \mathscr{A}_p(v)$ for all $g \in G,\ v \in T_p P;$ (9.74)

(d) $\mathscr{A}$ is smooth.

Property (c) is analogous to (8.72) for the Maurer–Cartan form.

The following theorem allows us to identify a connection with its 1-form.

THEOREM

If $\mathscr{A} \in \Lambda^1(P, \mathfrak{g})$ enjoys the properties (b)–(d) above, there is a uniquely determined connection on P such that $\mathscr{A}$ is its 1-form. The horizontal tangent spaces are given as

$$H_p = \{v \in T_p P \,|\, \mathscr{A}_p(v) = 0\}. \qquad (9.75)$$

Note that the connection 1-forms make up an affine subspace of $\Lambda^1(P, \mathfrak{g})$.

Using in the case of a trivial bundle $P = M \times G$ the notation of equation (9.58), we have for any connection 1-form $\mathscr{A}$ on P the representation

$$\mathscr{A}_{(x,h)}(v_M, v_G) = \mathrm{Ad}(h^{-1}) A_x(v_M) + \zeta_h(v_G), \qquad (9.76)$$

where A is a $\mathfrak{g}$-valued 1-form on the base space M and ζ the Maurer–Cartan form on G. It is easy to check that $\mathscr{A}$ as defined by (9.76) indeed satisfies the requirements (b)–(d) above.

9.9 The exterior covariant derivative

Let (P, M, π) be a principal bundle with structure group G, equipped with a connection. A differential form on P is called vertical (or horizontal, respectively) if it vanishes whenever one of the vectors on which it is evaluated is horizontal (respectively vertical). Note that horizontality of forms is independent of the existence of a connection, whereas we need one for the definition of vertical forms.

For a q-form φ on P we define the exterior covariant derivative $\mathrm{D}\varphi$ to be a $(q+1)$-form given by

$$\mathrm{D}\varphi(v_1, v_2, \ldots, v_{q+1}) = \mathrm{d}\varphi(Hv_1, Hv_2, \ldots, Hv_{q+1}), \quad v_i \in T_p P. \qquad (9.77)$$

Remember that Hv is the horizontal part of the tangent vector v. Obviously, $\mathrm{D}\varphi$ is horizontal.

The operator D as just defined acts on forms on P, whereas the exterior covariant derivative introduced in chapter 4 was applied to forms on the base space. The relation between these two derivatives will be clarified in the next section. There we shall also explain how some of the identities below are connected with their similarly looking counterparts in chapter 4.

Describing the connection by its 1-form $\mathscr{A}$ we can define its curvature $\mathscr{F}$, a horizontal, $\mathfrak{g}$-valued 2-form on P:

$$\mathscr{F} := \mathrm{D}\mathscr{A}. \qquad (9.78)$$

(Remember that in chapter 4 working on the base space we could not write

F as the exterior covariant derivative of the connection 1-form A.) Moreover, $\mathscr{F}$ is equivariant; i.e.,

$$\tilde{R}_g^* \mathscr{F} = \mathrm{Ad}(g^{-1}) \mathscr{F} \tag{9.79}$$

for all $g \in G$. Using, in particular, (9.68) and (9.74), we can show this easily:

$$
\begin{aligned}
(\tilde{R}_g^* \mathscr{F})(v_1, v_2) &= \mathscr{F}(T\tilde{R}_g(v_1), T\tilde{R}_g(v_2)) = \mathrm{d}\mathscr{A}(HT\tilde{R}_g(v_1), HT\tilde{R}_g(v_2)) \\
&= \mathrm{d}\mathscr{A}(T\tilde{R}_g(Hv_1), T\tilde{R}_g(Hv_2)) = (\tilde{R}_g^* \, \mathrm{d}\mathscr{A})(Hv_1, Hv_2) \\
&= (\mathrm{d}\tilde{R}_g^* \mathscr{A})(Hv_1, Hv_2) = \mathrm{Ad}(g^{-1})(\mathrm{d}\mathscr{A})(Hv_1, Hv_2) \\
&= \mathrm{Ad}(g^{-1})(D\mathscr{A})(v_1, v_2) = \mathrm{Ad}(g^{-1})\mathscr{F}(v_1, v_2). \tag{9.80}
\end{aligned}
$$

The proof of Cartan's structure equation

$$\mathscr{F} = \mathrm{d}\mathscr{A} + \tfrac{1}{2}[\mathscr{A}, \mathscr{A}] \tag{9.81}$$

is more complicated and therefore omitted.

The Bianchi identity

$$D\mathscr{F} = 0 \tag{9.82}$$

follows easily from (9.81). For, let T_a denote a basis of $\mathfrak{g}$ and write $\mathscr{A}$ as

$$\mathscr{A} = \sum_a \psi^a T_a \tag{9.83}$$

with $\psi^a \in \Lambda^1 P$. Then (9.81) reads

$$\mathscr{F} = \sum_a \mathrm{d}\psi^a T_a + \tfrac{1}{2} \sum_{a,b} \psi^a \wedge \psi^b [T_a, T_b]. \tag{9.84}$$

So we get

$$
\begin{aligned}
D\mathscr{F}(v_1, v_2, v_3) &= \mathrm{d}\mathscr{F}(Hv_1, Hv_2, Hv_3) \\
&= \tfrac{1}{2} \sum_{a,b} ((\mathrm{d}\psi^a) \wedge \psi^b - \psi^a \wedge \mathrm{d}\psi^b)(Hv_1, Hv_2, Hv_3)[T_a, T_b], \tag{9.85}
\end{aligned}
$$

and this expression vanishes because $\mathscr{A}$ is vertical so that the forms ψ^a give zero on horizontal vectors.

Our last topic in this section is the structure equation for horizontal equivariant forms. Let ρ be a linear representation of the structure group G on the vector space V and $\tilde{\rho}$ the corresponding representation of $\mathfrak{g}$. A V-valued q-form φ on P such that

$$\tilde{R}_g^* \varphi = \rho(g^{-1}) \varphi \tag{9.86}$$

for all $g \in G$ is called equivariant. (In (9.79) we have already encountered a special case of this definition.) If φ is in addition horizontal, we have the structure equation

$$D\varphi = d\varphi + \tilde{\rho}(\mathscr{A}) \wedge \varphi. \tag{9.87}$$

For a matrix group G it is also written

$$D\varphi = d\varphi + \mathscr{A} \wedge \varphi. \tag{9.88}$$

In the special case where $V = \mathfrak{g}$, $\rho = \mathrm{Ad}$, $\tilde{\rho} = \mathrm{ad}$, we get

$$D\varphi = d\varphi + \mathrm{ad}(\mathscr{A}) \wedge \varphi = d\varphi + [\mathscr{A}, \varphi]. \tag{9.89}$$

It is easy to see that the exterior covariant derivative of an equivariant form is again equivariant (cf. (9.80)).

9.10 Projecting down to the base space

In this section we want to establish the connection between the formulas of the last section and the similar equations from chapter 4.

We consider a principal bundle (P, M, π) with structure group G, connection $\mathscr{A}$ and a family of local trivializations

$$\pi^{-1}(\mathscr{U}_r) \to \mathscr{U}_r \times G$$
$$\tag{9.90}$$
$$p \mapsto (\pi(p), f_r(p)).$$

These local trivializations determine local sections

$$\sigma_r \colon \mathscr{U}_r \to \pi^{-1}(\mathscr{U}_r) \tag{9.91}$$

given by

$$\sigma_r(x) = f_{r,x}^{-1}(e) \tag{9.92}$$

for $x \in \mathscr{U}_r$; i.e. under the local trivialization (9.90) $\sigma_r(x)$ is mapped onto (x, e): $f_r(\sigma_r(x)) = e$. Conversely, a family of local sections σ_r leads to local trivializations such that

$$f_r(\sigma_r(x)g) = g \tag{9.93}$$

for $g \in G$ (compare the proof of the theorem in section 9.7).

For $x \in \mathscr{U}_r \cap \mathscr{U}_s$ the two local sections σ_r and σ_s are related by the right action of the structure group:

$$\sigma_s(x) = \sigma_r(x)\gamma(x), \tag{9.94}$$

where $\gamma \colon \mathscr{U}_r \cap \mathscr{U}_s \to G$. It should not be too surprising that γ has something to do with the transition function g_{rs}:

$$g_{rs}(x)g = f_{r,x}(f_{s,x}^{-1}(g)) = f_{r,x}(\sigma_s(x)g) = f_{r,x}(\sigma_r(x)\gamma(x)g) = \gamma(x)g \qquad (9.95)$$

with $g \in G$. So we have $\gamma(x) = g_{rs}(x)$ and consequently

$$\sigma_s(x) = \sigma_r(x)g_{rs}(x). \qquad (9.96)$$

We use these local sections to pull back forms from P to the base space M. In particular, we may represent the connection with respect to the given local trivialization by the g-valued 1-form

$$A_r := \sigma_r^* \mathscr{A} \qquad (9.97)$$

on $\mathscr{U}_r$, which is identified with the gauge potential as defined in section 4.2.

One can show for $x \in \mathscr{U}_r \cap \mathscr{U}_s$:

$$A_{r,x} = \mathrm{Ad}(g_{sr}^{-1}(x))A_{s,x} + (g_{sr}^*\zeta)_x, \qquad (9.98)$$

where ζ is the Maurer–Cartan form on G. If G is a matrix group, this equation is often written as

$$A_r(v) = g_{sr}^{-1}(x)A_s(v)g_{sr}(x) + g_{sr}^{-1}(x)(\mathrm{d}g_{sr})_x(v), \quad v \in T_x M. \qquad (9.99)$$

This is just a gauge transformation on $\mathscr{U}_r \cap \mathscr{U}_s$ in the sense of chapter 4 (see (4.42)).

Having stepped down from P to M, we can go back by virtue of the following theorem.

THEOREM

Let $\{\mathscr{U}_r\}$ be an open covering of M. Given a family of local g-valued 1-forms $A_r \in \Lambda^1(\mathscr{U}_r, \mathfrak{g})$ which fulfil the compatibility condition (9.98) and a set of local sections $\sigma_r : \mathscr{U}_r \to \pi^{-1}(\mathscr{U}_r)$ satisfying (9.96), there is a unique connection $\mathscr{A}$ on P such that $A_r = \sigma_r^* \mathscr{A}$.

Pulling back the curvature $\mathscr{F} = \mathrm{D}\mathscr{A}$, we get local curvature (or field strength) 2-forms on M:

$$F_r := \sigma_r^* \mathscr{F}. \qquad (9.100)$$

Upon applying σ_r^* to Cartan's structure equation (9.81) we find

$$F_r = \sigma_r^*(\mathrm{d}\mathscr{A} + \tfrac{1}{2}[\mathscr{A}, \mathscr{A}]) = \mathrm{d}\sigma_r^*\mathscr{A} + \tfrac{1}{2}[\sigma_r^*\mathscr{A}, \sigma_r^*\mathscr{A}]$$

$$= \mathrm{d}A_r + \tfrac{1}{2}[A_r, A_r], \qquad (9.101)$$

an equation which was used as the definition of the field strength 2-form in chapter 4 (cf. (4.58)). For $x \in \mathscr{U}_r \cap \mathscr{U}_s$ we have

$$F_{r,x} = \mathrm{Ad}(g_{sr}^{-1}(x))F_{s,x}, \qquad (9.102)$$

in particular, for matrix groups:

$$F_r(v_1, v_2) = g_{sr}^{-1}(x)F_s(v_1, v_2)g_{sr}(x), \quad v_1, v_2 \in T_xM. \quad (9.103)$$

Finally, we want to project down a horizontal equivariant q-form φ on P with values in a vector space V. Applying the pullback by σ_r we get a V-valued q-form on $\mathcal{U}_r : \sigma_r^*\varphi$. For $x \in \mathcal{U}_r \cap \mathcal{U}_s$ one finds

$$\sigma_r^*\varphi|_x = \rho(g_{rs}(x))\sigma_s^*\varphi|_x, \quad (9.104)$$

where ρ is the same representation as in (9.86). Again, the transition from one trivializing neighbourhood to another is a gauge transformation in the sense of chapter 4.

Evaluating both sides of (9.104) on $u_1, u_2, \ldots, u_q \in \text{vect}(M)$, we see (cf. (9.46)) that the maps

$$\mathcal{U}_r \to V$$

$$x \mapsto (\sigma_r^*\varphi)_x(u_1(x), u_2(x), \ldots, u_q(x)) \quad (9.105)$$

define a section of an associated vector bundle with standard fibre V on which the structure group G acts according to the representation ρ. One describes this situation by saying that the local forms $\sigma_r^*\varphi$ fit together to make up a section valued q-form on M. As one can show, the correspondence between horizontal equivariant forms on P and section valued forms on M is one to one.

Note that all 0-forms, i.e. functions, on P are trivially horizontal. Furthermore, for $q = 0$ one can generalize the above correspondence replacing the vector space V by an arbitrary manifold F. So this extended version connects functions $\varphi: P \to F$ such that

$$\tilde{R}_g^*\varphi = \varphi \circ \tilde{R}_g = \rho(g^{-1})\varphi \quad (9.106)$$

for all $g \in G$ and sections of an associated fibre bundle over M with standard fibre F on which G is represented by ρ. The section corresponding to the function φ is described locally by the maps

$$\sigma_r^*\varphi = \varphi \circ \sigma_r : \mathcal{U}_r \to F. \quad (9.107)$$

If F is a vector space, we may consider (9.107) as representing a multiplet of scalar matter fields.

As an example with $q = 2$ we quote the curvature 2-form $\mathcal{F}$. The local forms $F_r = \sigma_r^*\mathcal{F}$ define a section valued 2-form on M, where the associated vector bundle has the standard fibre $\mathfrak{g}$ and the representation ρ of G is the adjoint representation.

The structure equation (9.87) for horizontal equivariant forms with values in a vector space V yields

$$\sigma_r^* \, \mathrm{D}\varphi = \mathrm{d}\sigma_r^* \, \varphi + \tilde{\rho}(A_r) \wedge \sigma_r^* \, \varphi. \qquad (9.108)$$

On the rhs we recognize the exterior covariant derivative of the V-valued q-form $\sigma_r^* \, \varphi$ in the sense of chapter 4. Thus we have established the relation between the exterior covariant derivatives on M and P as announced in the preceding section. In particular, the Bianchi identity (9.82), $\mathrm{D}\mathscr{F} = 0$, leads to the Bianchi identity (4.74):

$$\mathrm{d}F_r + [A_r, F_r] = 0. \qquad (9.109)$$

9.11 Geometry of the frame bundle

The frame bundle $F(M)$ of an n-dimensional manifold M is a particularly important principal bundle. A connection $\mathscr{A}$ on $F(M)$ is called an affine connection, the corresponding curvature is usually denoted by $\mathscr{R} = \mathrm{D}\mathscr{A}$. The forms $\mathscr{R}$ and $\mathscr{A}$ are gl_n-valued, i.e. matrix-valued.

On $F(M)$ we can define additional objects leading eventually to the notion of torsion. Let $p = (x, b_1, b_2, \ldots, b_n)$ be a point of $F(M)$; i.e., $b_1, b_2, \ldots, b_n$ is a basis of $T_x M$. Denoting the dual basis of $(T_x M)^*$ by $\beta^1, \beta^2, \ldots, \beta^n$ we define

$$\Phi_p : T_x M \to \mathbb{R}^n$$

$$(9.110)$$

$$v \mapsto (\beta^1(v), \beta^2(v), \ldots, \beta^n(v)).$$

By Φ_p each $v \in T_x M$ is mapped onto its expansion coefficients with respect to $b_1, b_2, \ldots, b_n$:

$$v = \sum_{i=1}^{n} \beta^i(v) b_i = \sum_{i=1}^{n} \Phi_p^i(v) b_i. \qquad (9.111)$$

We have

$$\Phi_{p\gamma} = \gamma^{-1} \Phi_p \qquad (9.112)$$

for all $\gamma \in GL_n$. This equation is proved as follows. We know from (9.33) that

$$p\gamma = \left(x, \sum_{i=1}^{n} \gamma^i_{\ 1} b_i, \sum_{i=1}^{n} \gamma^i_{\ 2} b_i, \ldots, \sum_{i=1}^{n} \gamma^i_{\ n} b_i \right). \qquad (9.113)$$

So we can write for $v \in T_x M$:

$$v = \sum_{i=1}^{n} \Phi_p^i(v) b_i = \sum_{i,j,k=1}^{n} \Phi_p^i(v)(\gamma^{-1})^k{}_i \gamma^j{}_k b_j$$

$$= \sum_{k=1}^{n} \left(\sum_{i=1}^{n} (\gamma^{-1})^k{}_i \Phi_p^i(v) \right) \left(\sum_{j=1}^{n} \gamma^j{}_k b_j \right)$$

$$= \sum_{k=1}^{n} \Phi_{p\gamma}^k(v) \left(\sum_{j=1}^{n} \gamma^j{}_k b_j \right). \quad \text{QED} \tag{9.114}$$

We use the maps Φ_p to define the so-called solder form ϑ. It is a 1-form on $F(M)$ with values in $\mathbb{R}^n$ given by

$$\vartheta_p(v) := \Phi_p(T_p \pi(v)), \tag{9.115}$$

where now $v \in T_p F(M)$. So we expand $T_p \pi(v) \in T_{\pi(p)} M$ with respect to the basis $b_1, b_2, \ldots, b_n$ contained in p and call the expansion coefficients $\vartheta_p^i(v)$:

$$T_p \pi(v) = \sum_{i=1}^{n} \vartheta_p^i(v) b_i. \tag{9.116}$$

Assuming an affine connection $\mathscr{A}$ to be given we define the torsion form $\mathscr{T}$, a 2-form on $F(M)$ with values in $\mathbb{R}^n$, as the exterior covariant derivative of ϑ:

$$\mathscr{T} := \mathrm{D}\vartheta. \tag{9.117}$$

Obviously, ϑ and $\mathscr{T}$ are horizontal. Furthermore, they are both equivariant:

$$\tilde{R}_\gamma^* \vartheta = \gamma^{-1} \vartheta, \quad \tilde{R}_\gamma^* \mathscr{T} = \gamma^{-1} \mathscr{T} \tag{9.118}$$

for all $\gamma \in GL_n$. For ϑ this is shown by the following calculation where $v \in T_p F(M)$:

$$(\tilde{R}_\gamma^* \vartheta)_p v = \vartheta_{p\gamma}(T_p \tilde{R}_\gamma(v)) = \Phi_{p\gamma}(T_{p\gamma} \pi(T_p \tilde{R}_\gamma(v)))$$

$$= \Phi_{p\gamma}(T_p(\pi \circ \tilde{R}_\gamma)(v)) = \Phi_{p\gamma}(T_p \pi(v))$$

$$= \gamma^{-1} \Phi_p(T_p \pi(v)) = \gamma^{-1} \vartheta_p(v). \tag{9.119}$$

The equivariance of $\mathscr{T}$ is now an immediate consequence of the above mentioned fact that the exterior covariant derivative of an equivariant form is equivariant.

We have the structure equations

$$\mathscr{R} = \mathrm{d}\mathscr{A} + \tfrac{1}{2}[\mathscr{A}, \mathscr{A}] \tag{9.120}$$

(cf. (9.81)) and

$$\mathscr{T} = \mathrm{d}\vartheta + \mathscr{A} \wedge \vartheta, \tag{9.121}$$

a special case of (9.88), as well as the two affine Bianchi identities

$$D\mathscr{R} = 0, \tag{9.122}$$

$$D\mathscr{T} = \mathscr{R} \wedge \vartheta. \tag{9.123}$$

Equation (9.122) is just (9.82) specialized to an affine connection. In order to prove (9.123) we start by applying d to (9.121):

$$d\mathscr{T} = d^2\vartheta + d(\mathscr{A} \wedge \vartheta) = (d\mathscr{A}) \wedge \vartheta - \mathscr{A} \wedge d\vartheta. \tag{9.124}$$

So we get for vectors v_1, v_2, v_3 tangent to $F(M)$:

$$D\mathscr{T}(v_1, v_2, v_3) = d\mathscr{T}(Hv_1, Hv_2, Hv_3)$$

$$= (d\mathscr{A}) \wedge \vartheta(Hv_1, Hv_2, Hv_3)$$

$$- \mathscr{A} \wedge d\vartheta(Hv_1, Hv_2, Hv_3). \tag{9.125}$$

Since $\mathscr{A}$ is vertical and ϑ is horizontal, we can write

$$D\mathscr{T}(v_1, v_2, v_3) = (d\mathscr{A}) \wedge \vartheta(Hv_1, Hv_2, Hv_3)$$

$$= (D\mathscr{A}) \wedge \vartheta(v_1, v_2, v_3); \tag{9.126}$$

i.e.

$$D\mathscr{T} = (D\mathscr{A}) \wedge \vartheta = \mathscr{R} \wedge \vartheta.$$

As in the case of a general principal bundle we may use local sections to project our forms down from $F(M)$ to the base space M. A local section σ of $F(M)$ over some open subset $\mathscr{U} \subset M$ is determined by an n-frame $b_1, b_2, \ldots, b_n$ on $\mathscr{U}$:

$$\sigma(x) = (x, b_1(x), b_2(x), \ldots, b_n(x)), \quad x \in \mathscr{U}. \tag{9.127}$$

We can express the pull back $\sigma^*\vartheta$ of the solder form in terms of the dual n-frame $\beta^1, \beta^2, \ldots, \beta^n$:

$$(\sigma^*\vartheta)_x(v) = \vartheta_{\sigma(x)}(T_x\sigma(v)) = \Phi_{\sigma(x)}(T_{\sigma(x)}\pi(T_x\sigma(v)))$$

$$= \Phi_{\sigma(x)}(T_x(\pi\circ\sigma)(v)) = \Phi_{\sigma(x)}(v)$$

$$= (\beta^1(v), \beta^2(v), \ldots, \beta^n(v)) \tag{9.128}$$

for $x \in \mathscr{U}$, $v \in T_xM$. So we get

$$\sigma^*\vartheta^i = \beta^i. \tag{9.129}$$

Applying σ^* to the structure equation (9.121), we find

$$\sigma^*\mathscr{T} = d\sigma^*\vartheta + (\sigma^*\mathscr{A}) \wedge (\sigma^*\vartheta).$$

We identify (for $n = 4$) $\sigma^*\mathscr{A}$ with the gl_4-valued connection 1-form Γ of chapter 5. So we can write

$$\sigma^* \mathcal{T}^i = \mathrm{d}\beta^i + \sum_j \Gamma^i{}_j \wedge \beta^i = T^i \tag{9.130}$$

according to the previous definition (5.12) of torsion. Since $\mathcal{T}$ is horizontal and equivariant, we may apply (9.88) and get from the Bianchi identity (9.123)

$$\sigma^* \mathrm{D}\mathcal{T} = \mathrm{d}\sigma^* \mathcal{T} + (\sigma^* \mathcal{A}) \wedge (\sigma^* \mathcal{T})$$

$$= (\sigma^* \mathcal{R}) \wedge (\sigma^* \vartheta). \tag{9.131}$$

Putting $\sigma^* \mathcal{R} = R$ we recover the Bianchi identity (5.16):

$$\mathrm{d}T + \Gamma \wedge T = R \wedge \beta. \tag{9.132}$$

9.12 Parallel translation

Let (P, M, π) be a principal bundle with structure group G and a connection $\mathcal{A}$. For a curve C in M given by a parameter representation

$$Q: [0, 1] \to M, \tag{9.133}$$

which has to be continuous but is allowed to be only piecewise smooth, we define a horizontal lift to be a curve $\tilde{C}$ in P described by a parameter representation

$$\tilde{Q}: [0, 1] \to P \tag{9.134}$$

such that:

(a) $\dot{\tilde{Q}}|_\tau \in H_{Q(\tau)}$ for $\tau \in [0, 1]$;
(b) $\pi \circ \tilde{Q} = Q$.

In words, a horizontal lift of C is a curve in P which is mapped onto C by the projection π and whose tangent vectors are all horizontal.

THEOREM

For each $p \in \pi^{-1}(Q(0))$ there is one and only one horizontal lift with $\tilde{Q}(0) = p$.

So we can define a map

$$T_C: \pi^{-1}(Q(0)) \to \pi^{-1}(Q(1)) \tag{9.135}$$

assigning to each $p \in \pi^{-1}(Q(0))$ the end point $\tilde{Q}(1)$ of the unique horizontal lift of C starting at $\tilde{Q}(0) = p$. This map T_C, which turns out to be a diffeomorphism, is called parallel translation (or parallel transport) along C. It satisfies

$$T_C \circ \tilde{R}_g = \tilde{R}_g \circ T_C \tag{9.136}$$

for all $g \in G$, and parallel translation along the composition $C_1 C_2$ of the curves C_1 and C_2 with $Q_1(1) = Q_2(0)$ is given by

$$T_{C_1 C_2} = T_{C_2} \circ T_{C_1}. \tag{9.137}$$

Parallel transport in the principal bundle P gives rise to the notion of parallel transport in associated vector bundles. Let (E, M, π_E) be a vector bundle associated to P via the representation ρ of G on the standard fibre V. By $\tilde{\rho}$ we denote the corresponding representation of g. For simplicity we assume that C is contained in one trivializing neighbourhood $\mathcal{U}$. (If this is not the case, we subdivide C into sufficiently small pieces and consider each of them separately.) Let

$$\pi^{-1}(\mathcal{U}) \to \mathcal{U} \times G$$
$$p \mapsto (\pi(p), f(p)), \tag{9.138}$$

$$\pi_E^{-1}(\mathcal{U}) \to \mathcal{U} \times V$$
$$b \mapsto (\pi_E(b), f_E(b)) \tag{9.139}$$

be local trivializations of P and E, respectively. Parallel transport of an initial vector $b(0) \in \pi_E^{-1}(Q(0))$ along C leads to vectors $b(\tau) \in \pi_E^{-1}(Q(\tau))$ defined such that

$$f_E(b(\tau)) = \rho(f(\tilde{Q}(\tau)) f(\tilde{Q}(0))^{-1}) f_E(b(0)), \tag{9.140}$$

where $\tilde{Q}$ represents a horizontal lift of C. One can show that this definition of $b(\tau)$ is indeed independent of the choice of the horizontal lift and the local trivializations.

If $p \in \pi^{-1}(Q(0))$ is represented by $(Q(0), g)$ relative to the trivialization (9.138), we find that $T_C(p)$ corresponds to $(Q(1), \Gamma[C]g)$ with a $\Gamma[C] \in G$; i.e. $f(T_C(p)) = \Gamma[C]g = \Gamma[C]f(p)$. Consequently,

$$f_E(b(1)) = \rho(\Gamma[C]) f_E(b(0)), \tag{9.141}$$

and we can identify $\Gamma[C]$ with the parallel transporter introduced in section 4.2.

By means of the section

$$\sigma: \mathcal{U} \to \pi^{-1}(\mathcal{U}) \tag{9.142}$$

associated with the local trivialization (9.138) ($f(\sigma(x)) = e$ for all $x \in \mathcal{U}$) we pull back $\mathcal{A}$ to the base space, with the result $A = \sigma^* \mathcal{A}$, and from

(9.140) one may deduce the differential equation (4.38) for parallel transport along C:

$$\frac{\mathrm{d}}{\mathrm{d}\tau} f_E(b(\tau)) = - \tilde{\rho}(A(\dot{Q}(\tau))) f_E(b(\tau)). \qquad (9.143)$$

In the case of a Riemannian manifold M, an affine connection is called metric if the corresponding parallel transport in TM preserves the metric (cf. section 5.3). For each Riemannian manifold there is one and only one metric connection with vanishing torsion, the Riemannian connection. Locally it is described by the formulas of chapter 5.

9.13 The holonomy group

The path dependence of parallel translation gives rise to the notion of the holonomy group. Let (P, M, π) be a principal bundle endowed with a connection $\mathcal{A}$. As usual, we denote the structure group by G. Parallel transport of $p \in P$ along a closed curve C leads back to the fibre in which p lies. Consequently there must exist a uniquely determined $g \in G$ such that $T_C(p) = pg$. As C varies over all closed paths through $\pi(p)$, the corresponding gs form a subgroup of G called the holonomy group $\mathcal{H}(p)$ of $\mathcal{A}$ at p. Due to (9.136) we have

$$\mathcal{H}(pg) = g^{-1} \mathcal{H}(p)g \qquad (9.144)$$

for all $g \in G$. Furthermore,

$$\mathcal{H}(p) = \mathcal{H}(T_C(p)) \qquad (9.145)$$

for any (not necessarily closed) curve C in M. If M is connected, one can show that $\mathcal{H}(p)$ is a Lie subgroup of G.

As an example consider $P = F(S^2)$ with the Riemannian connection belonging to the standard metric on S^2. The holonomy group at any point is $SO(2)$. One may visualize its elements as the rotations resulting from parallel transport of tangent vectors along closed paths.

Remembering our discussion of curvature (field strength) in section 4.3, one should expect a connection between $\mathcal{F} = D\mathcal{A}$ and the holonomy groups. Such a connection is established by the following theorem of Ambrose and Singer.

THEOREM

Let M be connected. For $p_0 \in P$ the Lie algebra of $\mathcal{H}(p_0)$ is the subspace of g spanned by all elements $\mathcal{F}_p(v_1, v_2)$, where $v_1, v_2 \in T_p P$ and p can be reached from p_0 by parallel translation.

9.14 Gauge transformations

In chapters 4 and 5 a gauge transformation was a map $\gamma: \mathcal{U} \to G$ from spacetime, an open subset of $\mathbb{R}^4$, into a Lie group G. It acted in a well-defined way on the connection, the field strength, etc. Defining multiplication of such maps pointwise we arrived at the gauge group $^\# G$. In this section we want to generalize the concept of gauge transformation to nontrivial bundles, i.e. to a situation where, for instance, the connection can only locally be described by a g-valued 1-form on the base space.

For a principal bundle (P, M, π) with structure group G, local trivializations determine a family of local sections

$$\sigma_r: \mathcal{U}_r \to \pi^{-1}(\mathcal{U}_r) \tag{9.146}$$

such that

$$\sigma_r(x) = \sigma_s(x) g_{sr}(x), \quad x \in \mathcal{U}_r \cap \mathcal{U}_s, \tag{9.147}$$

with the transition functions

$$g_{sr}: \mathcal{U}_r \cap \mathcal{U}_s \to G \tag{9.148}$$

(cf. section 9.10). A connection $\mathcal{A}$ on P can be represented by the 1-forms $A_r = \sigma_r^* \mathcal{A}$ on the trivializing neighbourhoods $\mathcal{U}_r$. With the help of the Maurer–Cartan form ζ on G the relation between A_r and A_s is written as

$$A_{r,x} = \mathrm{Ad}\,(g_{sr}^{-1}(x)) A_{s,x} + (g_{sr}^* \zeta)_x. \tag{9.149}$$

So on $\mathcal{U}_r \cap \mathcal{U}_s$ the local potentials A_r and A_s are connected by a gauge transformation in the sense of chapter 4 (cf. (9.98), (9.99)). However, these transformations do not give rise to a globally defined transformation. Rather they appear as compatibility conditions for the local potentials A_r.

Now we ask the question: What happens if we replace the local sections σ_r by new local sections

$$\sigma_r': \mathcal{U}_r \to \pi^{-1}(\mathcal{U}_r)$$

$$x \mapsto \sigma_r'(x) = \sigma_r(x) \gamma_r^{-1}(x) \tag{9.150}$$

with $\gamma_r: \mathcal{U}_r \to G$? We get new 1-forms $A_r' = \sigma_r'^* \mathcal{A}$ representing the connection on $\mathcal{U}_r$, which are related to the old ones by

$$A_{r,x}' = \mathrm{Ad}\,(\gamma_r(x)) A_{r,x} + (\gamma_r^{-1})^* \zeta|_x. \tag{9.151}$$

Of course, the A'_r satisfy compatibility conditions analogous to (9.149):

$$A'_{r,x} = \mathrm{Ad}\,(g'^{-1}_{sr}(x))A'_{s,x} + (g'^{*}_{sr}\zeta)_x, \tag{9.152}$$

where the g'_{sr} are the transition functions which belong to the local trivializations defined by the sections σ'_r. These new transition functions are easily calculated:

$$\sigma'_r(x) = \sigma_r(x)\gamma_r^{-1}(x) = \sigma_s(x)g_{sr}(x)\gamma_r^{-1}(x) = \sigma'_s(x)\gamma_s(x)g_{sr}(x)\gamma_r^{-1}(x)$$

$$= \sigma'_s(x)g'_{sr}(x);$$

i.e.

$$g'_{sr}(x) = \gamma_s(x)g_{sr}(x)\gamma_r^{-1}(x), \quad x \in \mathcal{U}_r \cap \mathcal{U}_s. \tag{9.153}$$

The transformation (9.151) leading from A_r to A'_r is already defined on all of $\mathcal{U}_r$. But in order to get a globally defined transformation, we have to glue together the γ_r somehow; i.e., we must relate γ_r to γ_s on $\mathcal{U}_r \cap \mathcal{U}_s$. The following observation gives us a hint how this should be done: The compatibility conditions among the local gauge potentials A'_r are identical to those fulfilled by the original potentials A_r if

$$g'_{sr} = g_{sr}, \tag{9.154}$$

i.e. if

$$\gamma_r(x) = g_{sr}^{-1}(x)\gamma_s(x)g_{sr}(x)$$

$$= g_{rs}(x)\gamma_s(x)g_{rs}^{-1}(x). \tag{9.155}$$

The requirement (9.154) ensures that the transformation leading from the family of 1-forms $\{A_r\}$ to the new family $\{A'_r\}$ is a global object: Due to (9.155) the G-valued functions γ_r describe a section of a bundle B associated to P. This associated bundle has standard fibre G, and the action ρ^B of the structure group G on the standard fibre is given by inner automorphisms:

$$\rho_h^B g = hgh^{-1} \tag{9.156}$$

for $g, h \in G$. Note that B is not a principal bundle because the structure group is not represented by left translation on the standard fibre. Moreover, B admits global sections even if it is not trivial.

To proceed further we change our point of view. Up to now we kept $\mathscr{A}$ fixed and studied the variation of the local potentials A_r under a change of the trivializing sections. In the following we fix the A_r and transform instead $\mathscr{A}$ into $\mathscr{A}'$ such that

$$A_r = \sigma_r^* \mathscr{A} = \sigma_r'^* \mathscr{A}'. \tag{9.157}$$

The existence of uniquely determined connection 1-forms $\mathscr{A}$ and $\mathscr{A}'$ on P satisfying (9.157) is guaranteed by the reconstruction theorem of section 9.10, because the transition functions with respect to σ_r as well as those with respect to σ_r' are given by the functions g_{sr} appearing in the compatibility conditions (9.149).

How are $\mathscr{A}$ and $\mathscr{A}'$ related? According to the discussion in section 9.10, the section of B represented locally by $\gamma_r: \mathscr{U}_r \to G$ corresponds to a function $\Gamma: P \to G$ satisfying

$$\Gamma(pg) = g^{-1}\Gamma(p)g \qquad (9.158)$$

for $p \in P$, $g \in G$, and the functions γ_r can be recovered from Γ by means of the formula

$$\gamma_r = \Gamma \circ \sigma_r. \qquad (9.159)$$

Defining a map $f: P \to P$ by

$$f(p) := p\Gamma(p), \quad p \in P, \qquad (9.160)$$

we find for $g \in G$:

$$f(pg) = pg\Gamma(pg) = p\Gamma(p)g = f(p)g;$$

i.e.

$$f \circ \tilde{R}_g = \tilde{R}_g \circ f. \qquad (9.161)$$

So $(f, \mathrm{id}_G, \mathrm{id}_M)$ is a bundle isomorphism of (P, M, π) onto itself, a bundle automorphism. The pull back by f maps $\mathscr{A}$ onto another connection 1-form. One gets

$$f^*\mathscr{A}|_p = \mathrm{Ad}\,(\Gamma^{-1}(p))\mathscr{A}|_p + \Gamma^*\zeta\,|_p \qquad (9.162)$$

and for the local potential on $\mathscr{U}_r$:

$$\sigma_r^*(f^*\mathscr{A})|_x = \mathrm{Ad}\,(\gamma_r^{-1}(x))\sigma_r^*\mathscr{A}|_x + \gamma_r^*\zeta|_x. \qquad (9.163)$$

Representing the connection $f^*\mathscr{A}$ on M by means of new local sections σ_r', where

$$\sigma_r'(x) = \sigma_r(x)\gamma_r^{-1}(x), \quad x \in \mathscr{U}_r, \qquad (9.164)$$

we find (cf. (9.151)):

$$\begin{aligned}
\sigma_r'^*(f^*\mathscr{A})|_x &= \mathrm{Ad}\,(\gamma_r(x))\sigma_r^*(f^*\mathscr{A})|_x + (\gamma_r^{-1})^*\zeta|_x \\
&= \mathrm{Ad}\,(\gamma_r(x))\,\mathrm{Ad}\,(\gamma_r^{-1}(x))(\sigma_r^*\mathscr{A})|_x \\
&\quad + \mathrm{Ad}\,(\gamma_r(x))\gamma_r^*\zeta|_x + (\gamma_r^{-1})^*\zeta|_x \\
&= \sigma_r^*\mathscr{A}|_x. \qquad (9.165)
\end{aligned}$$

Comparing with (9.157) we see that the desired relation between $\mathscr{A}$ and $\mathscr{A}'$ is given by

$$\mathscr{A}' = f^*\mathscr{A}. \tag{9.166}$$

As (9.163) shows, the local potentials constructed from $\mathscr{A}$ and $\mathscr{A}'$ with the help of a fixed section σ_r are connected by a gauge transformation in the sense of chapter 4. Therefore we call a bundle automorphism of the form $(f, \mathrm{id}_G, \mathrm{id}_M)$, which is also named vertical automorphism, a gauge transformation. We have seen that it can be equivalently described by a G-valued function Γ on P satisfying (9.158) or by a section of the associated bundle B, which is given locally by the maps (9.159).

In the case of a trivial bundle $P = M \times G$ we have one map $\gamma: M \to G$ representing a gauge transformation, and one gets

$$\Gamma(x, g) = g^{-1}\gamma(x)g, \tag{9.167}$$

$$f(x, g) = (x, \gamma(x)g), \quad x \in M, \quad g \in G. \tag{9.168}$$

Returning to an arbitrary principal bundle we define the gauge group to be the set of all gauge transformations with the composition of the diffeomorphisms f as multiplication. How does this multiplication look in terms of the corresponding functions Γ? With

$$f_i(p) = p\Gamma_i(p), \quad i = 1, 2, \tag{9.169}$$

we find

$$(f_2 \circ f_1)(p) = f_2(p\Gamma_1(p)) = p\Gamma_1(p)\Gamma_2(p\Gamma_1(p))$$

$$= p\Gamma_2(p)\Gamma_1(p). \tag{9.170}$$

So the G-valued functions Γ are multiplied pointwise. The map

$$\gamma_r := (\Gamma_2\Gamma_1) \circ \sigma_r \tag{9.171}$$

representing the product $\Gamma_2\Gamma_1$ locally on $\mathscr{U}_r$ is easily expressed in terms of

$$\gamma_r^{(i)} := \Gamma_i \circ \sigma_r.$$

We get for $x \in \mathscr{U}_r$

$$\gamma_r(x) = \Gamma_2(\sigma_r(x))\Gamma_1(\sigma_r(x)) = \gamma_r^{(2)}(x)\gamma_r^{(1)}(x), \tag{9.172}$$

i.e. the product in the gauge group is given by pointwise multiplication of the maps γ_r, in accordance with our discussion in chapter 4.

How does a gauge transformation act on scalar matter fields or, more generally, on horizontal equivariant forms φ on P? We define the gauge transform of φ to be the horizontal equivariant form

$$\varphi' := f^* \varphi \tag{9.173}$$

with $f : P \to P$ as above. On $\mathcal{U}_r$ one finds

$$\sigma_r^* \varphi'|_x = \rho(\gamma_r^{-1}(x)) \sigma_r^* \varphi|_x, \tag{9.174}$$

where ρ is the representation of G appearing in the definition of equivariance (cf. (9.86)). So we recover locally the formulas of chapter 4 (see, e.g., (4.32), (4.63)).

Finally we want to show that diffeomorphisms of the base space M give rise to bundle automorphisms, too. For a trivial bundle $P = M \times G$ and $F \in \text{Diff}(M)$ one can define $f_P : P \to P$ by

$$f_P(x, g) := (F(x), g), \tag{9.175}$$

and one sees immediately that (f_P, id_G, F) is a bundle automorphism. However, if P is an arbitrary principal bundle over M, the construction we are going to describe will work only for diffeomorphisms F which can be smoothly deformed into the identity; i.e. we have to require $F \in (\text{Diff}(M))^e$. Let

$$\begin{aligned} h : [0, 1] \times M &\to M \\ (\tau, x) &\mapsto h_\tau(x) \end{aligned} \tag{9.176}$$

be such that

$$\begin{aligned} h_\tau &\in \text{Diff}(M), \\ h_0 &= \text{id}_M, \quad h_1 = F. \end{aligned} \tag{9.177}$$

For each $x \in M$ we get a curve C_x, represented by

$$\begin{aligned} [0, 1] &\to M \\ \tau &\mapsto h_\tau(x), \end{aligned} \tag{9.178}$$

which joins x and $F(x)$. Now choose a fixed connection $\mathcal{A}^\circ$ on P. For $p \in \pi^{-1}(x)$ we define $f_P(p) \in \pi^{-1}(F(x))$ to be the parallel transport of p along C_x with respect to $\mathcal{A}^\circ$. Note that the diffeomorphism $f_P : P \to P$ depends on the choice of h and $\mathcal{A}^\circ$. Due to (9.136) it satisfies

$$\tilde{R}_g f_P(p) = f_P(\tilde{R}_g p), \quad g \in G, \quad p \in P. \tag{9.179}$$

Therefore, (f_P, id_G, F) is a bundle automorphism. These automorphisms do not in general commute with automorphisms of the form $(f, \text{id}_G, \text{id}_M)$, i.e. gauge transformations. Formulas for the infinitesimal version will be given in chapter 12.

Problems

9.1 Give a differentiable atlas for the frame bundle $F(M)$.

9.2 Consider $SU(2)$ as the subgroup of $SU(3)$ consisting of all $SU(3)$-matrices of the form

$$\begin{pmatrix} 1 & 0 & 0 \\ 0 & & \\ 0 & & * \end{pmatrix}$$

Then, according to the theorem mentioned at the end of section 9.5, $SU(3)$ may be regarded as an $SU(2)$-bundle over S^5. Find local trivializations of this bundle over $S^5 - \{$north pole$\}$ and $S^5 - \{$south pole$\}$. Calculate the corresponding transition function.

9.3 Show that a vector bundle with an m-dimensional fibre is trivial if and only if it admits m sections linearly independent at each point.

10

Monopoles, instantons, and related fibre bundles

In this chapter we shall discuss special solutions to field equations of gauge theories and related fibre bundles. In doing so we shall encounter concrete examples of some of the abstract concepts introduced in the previous chapter.

10.1 The Dirac monopole

In order to describe a magnetic monopole within electrodynamics, Dirac (1931) used the vector potential

$$A = i\frac{m}{2r}\frac{1}{x^3 - r}(x^1 dx^2 - x^2 dx^1). \tag{10.1}$$

Remember that, according to our conventions, the 'physical' potential is given by $(i/g)A$, where g is the coupling constant, i.e. the electric charge. We work in Minkowski space with Cartesian coordinates x^0, x^1, x^2, x^3, and r denotes the distance from the origin in 3-space:

$$r^2 = (x^1)^2 + (x^2)^2 + (x^3)^2. \tag{10.2}$$

At present, m is an arbitrary real constant.

A strange feature of the potential (10.1) is that it is singular on the positive x^3-axis. Nevertheless, the corresponding field strength

$$F = dA = i\frac{m}{2r^3}(x^1 dx^2 \wedge dx^3 + x^2 dx^3 \wedge dx^1 + x^3 dx^1 \wedge dx^2) \tag{10.3}$$

has a singularity only at $r = 0$, which is unavoidable for a point source. Comparing the 'physical' field strength $(i/g)F$ with (4.4), we see that the electric field vanishes and that the magnetic induction is given by

$$B_j = -\frac{m}{2gr^3}x^j, \quad j = 1, 2, 3. \tag{10.4}$$

So (10.3) is indeed the field strength of a magnetic monopole with magnetic charge $(-m)/(2g)$ located at the origin. For $r \neq 0$ Maxwell's equations are satisfied:

$$dF = 2\pi i m \delta(x^1)\delta(x^2)\delta(x^3)\,dx^1 \wedge dx^2 \wedge dx^3,$$
$$\delta F = 0. \tag{10.5}$$

Next we want to study the behaviour of the 'string' singularity in A under a gauge transformation

$$A \to A' = A + \gamma^{-1}\,d\gamma, \tag{10.6}$$

where γ is a $U(1)$-valued function on spacetime. For convenience, we introduce polar coordinates r, ϑ, φ (cf. equation (3.27)), and get

$$A = \tfrac{1}{2}im(-1 - \cos\vartheta)\,d\varphi. \tag{10.7}$$

Choosing

$$\gamma(x) = e^{im\varphi} \tag{10.8}$$

we obtain

$$A' = \tfrac{1}{2}im(1 - \cos\vartheta)\,d\varphi = i\frac{m}{2r}\frac{1}{x^3 + r}(x^1\,dx^2 - x^2\,dx^1). \tag{10.9}$$

So our gauge transformation has shifted the singularity from the positive to the negative x^3-axis. Note, however, that γ is singular along the whole x^3-axis, because there the azimuthal angle φ is not defined. Furthermore, γ is single-valued only if m is an integer: The string singularity can be moved around by gauge transformations and is therefore unobservable if and only if $m \in \mathbb{Z}$. This is Dirac's 'quantization condition'.

Since in the expression for A and A' in terms of polar coordinates the radius r does not appear, we may consider our potentials as living on the 2-sphere S^2: On $\mathcal{U}_1 = S^2 - \{(0, 0, -1)\}$ we have the potential

$$A_1 = \tfrac{1}{2}im(1 - \cos\vartheta)\,d\varphi \tag{10.10}$$

and on $_2 = S^2 - \{(0, 0, 1)\}$,

$$A_2 = \tfrac{1}{2}im(-1 - \cos\vartheta)\,d\varphi. \tag{10.11}$$

On $\mathcal{U}_1 \cap \mathcal{U}_2$ the potentials are connected by a gauge transformation:

$$A_1 = A_2 + \gamma^{-1}\,d\gamma. \tag{10.12}$$

According to the theorem in section 9.10, A_1 and A_2 determine a connection on a $U(1)$ principal bundle over S^2 whose transition function $g_{21}: \mathcal{U}_1 \cap \mathcal{U}_2 \to U(1)$ is given by γ. For different values of the integer m these bundles are not isomorphic. In particular, if $m \neq 0$, the bundle is nontrivial. Thus we can avoid singular potentials and singular gauge transformations by admitting nontrivial fibre bundles and considering the potentials only on suitable neighbourhoods. For the case $m = 1$, details will be given in section 10.5.

10.2 The 't Hooft–Polyakov monopole

The 't Hooft–Polyakov monopole ('t Hooft 1974; Polyakov 1974) is a static finite energy solution to the field equations of an $SO(3)$ gauge theory with a triplet of real scalar fields transforming according to the fundamental representation of $SO(3)$. In this representation we have the generators

$$(T_a)^i{}_j = -\varepsilon_{aij}, \quad a, i, j \in \{1, 2, 3\} \tag{10.13}$$

(cf. (8.53)). The Lagrangian of our theory is given by (cf. (4.83))

$$\mathcal{L} = \left[-\frac{1}{4g^2} \sum_{a=1}^{3} F_{\mu\nu}^a F^{a,\mu\nu} + \tfrac{1}{2} \sum_{i=1}^{3} (D_\mu \Phi)^i (D^\mu \Phi)^i - U(\Phi) \right]$$

$$\times dx^0 \wedge dx^1 \wedge dx^2 \wedge dx^3 \tag{10.14}$$

with the 'potential'

$$U(\Phi) = \frac{\lambda}{4} \left(\frac{M^2}{\lambda} - \sum_{i=1}^{3} \Phi^i \Phi^i \right)^2, \tag{10.15}$$

where g is the gauge coupling, M a mass parameter and λ the scalar self-coupling. (The factor $\tfrac{1}{2}$ in front of the second term of $\mathcal{L}$ is due to the fact that here the scalar fields are real whereas in (4.83) they were assumed complex.) For a field configuration of finite energy, the contribution of the potential term U has to vanish at spatial infinity. So we must require a boundary condition of the form

$$\Phi^i \to \frac{M}{\sqrt{\lambda}} n^i \quad \text{as} \quad r = \left(\sum_{i=1}^{3} (x^i)^2 \right)^{1/2} \to \infty \tag{10.16}$$

with

$$\sum_{i=1}^{3} (n^i(x))^2 = 1.$$

This amounts to a violation of gauge invariance which is analogous to spontaneous symmetry breaking in quantum field theory.

To get the 't Hooft–Polyakov monopole we make the following static ansatz where we express the gauge potential A in terms of the matrices

T_a given in (10.13):

$$g\Phi^i(x) = \frac{x^i}{r^2} f(r), \quad i = 1, 2, 3,$$

$$A = \frac{1 - h(r)}{r^2} \sum_{a,i,j=1}^{3} T_a \varepsilon_{aij} x^j dx^i. \tag{10.17}$$

Note the characteristic 'hedgehog shape' of the configuration of the scalar fields: The direction of the vector (Φ^1, Φ^2, Φ^3) is identified with the radial direction in 3-space. So the $SO(3)$ symmetry is broken down to the $U(1)$ invariance under rotations about the radial direction (cf. (10.63) below). This remaining $U(1)$ is to be identified with the $U(1)$ of electromagnetism.

The field equations imply the following system of ordinary nonlinear differential equations for f and h:

$$r^2 \frac{d^2 f}{dr^2} = f\left(2h^2 - M^2 r^2 + \frac{\lambda}{g^2} f^2\right),$$

$$r^2 \frac{d^2 h}{dr^2} = h(h^2 - 1 + f^2). \tag{10.18}$$

The boundary conditions are

$$h(r) \to 0, \quad f(r)/r \to \beta := gM/\sqrt{\lambda} \quad \text{as} \quad r \to \infty,$$

$$h(0) = 1, \quad f(0) = 0, \tag{10.19}$$

and one finds

$$h(r) \to c_1 r e^{-\beta r},$$

$$f(r) \to \beta r + c_2 e^{-M\sqrt{2}r} \tag{10.20}$$

as $r \to \infty$, whereas for $r \to 0$

$$h(r) \to 1 + gc_3 r^2,$$

$$f(r) \to gc_4 r^2. \tag{10.21}$$

Here $c_1, \ldots, c_4$ are constants, which may depend on M, λ, and g. In particular, (10.21) shows that the fields A and Φ are not singular at the origin.

In the limit $M^2 \to 0$, $\lambda \to 0$, and M^2/λ fixed, Prasad & Sommerfield (1975) and Bogomolnyi (1976) found a solution to (10.18):

$$h(r) = \beta r / \sinh(\beta r), \tag{10.22}$$

$$f(r) = -1 + \beta r \cosh(\beta r) / \sinh(\beta r).$$

How is the 't Hooft–Polyakov monopole related to the Dirac monopole? To answer this question we perform a gauge transformation

$$A \to A' = \gamma A \gamma^{-1} + \gamma \, d\gamma^{-1},$$

$$\Phi \to \Phi' = \gamma \Phi \tag{10.23}$$

with an $SO(3)$-valued function γ. We choose

$$\gamma(x) = R\left(\begin{pmatrix} \cos\frac{1}{2}\vartheta & e^{-i\varphi}\sin\frac{1}{2}\vartheta \\ -e^{i\varphi}\sin\frac{1}{2}\vartheta & \cos\frac{1}{2}\vartheta \end{pmatrix}\right). \tag{10.24}$$

Here ϑ, φ are polar coordinates and $R: SU(2) \to SO(3)$ is the homomorphism constructed in section 8.5 (cf. (8.58)) where it was called φ. So we rotate the scalar field, which initially points into the radial direction, such that it becomes parallel to the x^3-axis:

$$g\Phi'^i = \delta_{i3} f(r)/r,$$

$$A'^1 = -h(r)(\sin\varphi \, d\vartheta + \cos\varphi \sin\vartheta \, d\varphi),$$

$$A'^2 = h(r)(\cos\varphi \, d\vartheta - \sin\varphi \sin\vartheta \, d\varphi), \tag{10.25}$$

$$A'^3 = -(1 - \cos\vartheta)\,d\varphi.$$

Note that γ and A'^3 are singular along the negative x^3-axis. As $r \to \infty, h$ vanishes exponentially, and only the third component of the gauge field survives. Comparing the 'physical' version of this potential,

$$-\frac{1}{g}(1 - \cos\vartheta)\,d\varphi, \tag{10.26}$$

with the 'physical' potential of the Dirac monopole (cf. (10.10)),

$$-\frac{m}{2g}(1 - \cos\vartheta)\,d\varphi, \tag{10.27}$$

and identifying the gs, we conclude that the 't Hooft–Polyakov monopole seen from afar looks like a Dirac monopole with $m = 2$.

From the viewpoint of fibre bundle theory one faces the following problem. Since the fields in the 't Hooft–Polyakov monopole are regular everywhere, they come from a trivial bundle, whereas the potential of the Dirac monopole (with $m \neq 0$) possessing a string singularity determines a nontrivial principal bundle. How is it then possible to recover the Dirac monopole from the 't Hooft–Polyakov monopole? With the help of the above gauge transformation we got the string singularity by 'brute force' admitting a singular γ. The answer within the framework of fibre bundles will be given in section 10.6.

10.3 The Yang–Mills instanton

We consider a pure Yang–Mills theory based upon the symmetry group $G = SU(2)$ in $\mathbb{R}^4$ with flat Euclidean metric and look for finite action solutions of the field equation (4.86). As already mentioned in section 4.5, the Bianchi identity (4.74) implies that (anti-) self-dual fields automatically satisfy the field equation. Therefore we may solve (4.86) by finding a solution of the (anti-) self-duality condition $F = \pm *F$, a system of first-order differential equations for the gauge potential A.

We make an ansatz which differs from a pure gauge only by a function of

$$r := \left[\sum_{i=1}^{4} (x^i)^2 \right]^{1/2}, \tag{10.28}$$

where $x^1, \ldots, x^4$ are Cartesian coordinates; i.e. we set

$$A_x = f(r)\gamma^* \zeta|_x = f(r)\gamma^{-1}\, d\gamma|_x \tag{10.29}$$

with an $SU(2)$-valued function γ and the Maurer–Cartan form ζ on $SU(2)$. If f tends to 1 as $r \to \infty$, the potential becomes a pure gauge at infinity. Consequently, the field strength F vanishes there, and the action is finite, provided the asymptotic values are reached sufficiently fast.

Choose

$$\gamma(x) = \frac{1}{r}\left(x^4 + 2 \sum_{j=1}^{3} \tau_j x^j \right)$$

$$= \frac{1}{r}\begin{pmatrix} x^4 - ix^3 & -x^2 - ix^1 \\ x^2 - ix^1 & x^4 + ix^3 \end{pmatrix} \tag{10.30}$$

with $\tau_j = -\tfrac{1}{2}i\sigma_j$ (cf. (8.55)). This map, which is singular at $r = 0$, identifies the 3-sphere of radius r with $SU(2)$ (cf. (8.11)). So, $\gamma^{-1}\, d\gamma$ differs from ζ only by this identification (for fixed r). We calculate

$$\gamma^* \zeta = \gamma^{-1}\, d\gamma = \sum_{j=1}^{3} \tau_j \tilde{\zeta}^j \tag{10.31}$$

with

$$\tilde{\zeta}^j = -\frac{2}{r^2}\left(x^j\, dx^4 - x^4\, dx^j + \sum_{k,l=1}^{3} \varepsilon_{jkl} x^k\, dx^l \right) = \gamma^* \zeta^j, \quad j = 1, 2, 3, \tag{10.32}$$

the pull back of the Maurer–Cartan forms ζ^j. Using the Maurer–Cartan structure equation (8.73) we find

$$F = dA + A \wedge A$$

$$= \sum_{l=1}^{3} \tau_l \left[\frac{df}{dr} dr \wedge \tilde{\zeta}^l + \tfrac{1}{2}(f^2 - f) \sum_{j,k=1}^{3} \varepsilon_{jkl} \tilde{\zeta}^j \wedge \tilde{\zeta}^k \right], \qquad (10.33)$$

where

$$dr = \sum_{j=1}^{4} \frac{x^j}{r} dx^j.$$

Since we need $*F$, it is advantageous to express F in terms of an orthonormal frame. One easily verifies that $e^1, \ldots, e^4$ with

$$e^j := \tfrac{1}{2} r \tilde{\zeta}^j, \quad j = 1, 2, 3, \quad e^4 := dr \qquad (10.34)$$

is an oriented orthonormal frame on $\mathbb{R}^4 - \{0\}$. We can write

$$F = \sum_{l=1}^{3} \tau_l \left[\frac{2}{r} \frac{df}{dr} e^4 \wedge e^l + \frac{2}{r^2}(f^2 - f) \sum_{j,k=1}^{3} \varepsilon_{jkl} e^j \wedge e^k \right], \qquad (10.35)$$

and applying the definition (3.16) of the Hodge star we get

$$*F = \sum_{l=1}^{3} \tau_l \left[-\frac{4}{r^2}(f^2 - f) e^4 \wedge e^l - \frac{1}{r} \frac{df}{dr} \sum_{j,k=1}^{3} \varepsilon_{jkl} e^j \wedge e^k \right]. \qquad (10.36)$$

Hence, the self-duality condition $F = *F$ leads to the following differential equation for f:

$$\frac{df}{dr} = -\frac{2}{r} f(f - 1). \qquad (10.37)$$

(The anti-self-dual case can be treated analogously.) Since the solution

$$f(r) = r^2/(r^2 + c^2), \quad c \in \mathbb{R}, \qquad (10.38)$$

vanishes at $r = 0$, the resulting so-called instanton potential

$$A = \frac{r^2}{r^2 + c^2} \gamma^{-1} d\gamma \qquad (10.39)$$

(Belavin et al. 1975) is regular on all of $\mathbb{R}^4$. Furthermore, f tends to 1 as $r \to \infty$, and the action is indeed finite:

$$\frac{1}{g^2} \int \mathrm{tr}(F \wedge *F) = -\frac{8\pi^2}{g^2}. \qquad (10.40)$$

How this solution of the Yang–Mills equation gives rise to an $SU(2)$ bundle over S^4 will be explained in section 10.7.

10.4 A gravitational instanton

Following Eguchi & Hanson (1979) we try to imitate the construction of the Yang–Mills instanton solution in the case of gravity. So we look for a metric of Euclidean signature $(\eta_{ij} = \delta_{ij})$ described locally by an orthonormal frame $e^1, \ldots, e^4$. The equation we have to solve is Einstein's equation (5.31) without matter,

$$\sum_{a,b,d=1}^{4} R^{ab} \wedge e^d \varepsilon_{abcd} = 0, \quad c = 1, \ldots, 4, \qquad (10.41)$$

where R is the curvature 2-form. Note that R as well as the spin connection ω take their values in the Lie algebra of real, antisymmetric 4×4 matrices, because we have assumed Euclidean signature. Defining the dual curvature $\tilde{R}$ by

$$\tilde{R}^{ab} := \tfrac{1}{2} \sum_{c,d=1}^{4} \varepsilon_{abcd} R^{cd} \qquad (10.42)$$

we can rewrite (10.41) as

$$\tilde{R} \wedge e = 0. \qquad (10.43)$$

Consequently, if the curvature is (anti-) self-dual,

$$\tilde{R} = \pm R, \qquad (10.44)$$

the field equation (10.43) is automatically satisfied due to the Bianchi identity $R \wedge e = 0$ (cf. (5.16)). This is closely analogous to the Yang–Mills case. However, (10.44) is a second-order differential equation for our fundamental quantities e^a, whereas the self-duality condition for Yang–Mills fields is a first-order differential equation for the potential A.

In order to find a differential equation of first order for e^a, we have to go one step further and define also a dual connection $\tilde{\omega}$:

$$\tilde{\omega}^{ab} := \tfrac{1}{2} \sum_{c,d=1}^{4} \varepsilon_{abcd} \omega^{cd}. \qquad (10.45)$$

From $R = d\omega + \omega \wedge \omega$ one can deduce:

$$\tilde{R} = d\tilde{\omega} + \tfrac{1}{2}(\omega \wedge \tilde{\omega} + \tilde{\omega} \wedge \omega). \qquad (10.46)$$

Therefore the (anti-) self-duality condition for ω, $\tilde{\omega} = \pm \omega$, a first-order differential equation for e^a, implies (10.44) and, a fortiori, the field equation (10.43).

We now choose an ansatz which differs from the flat metric only by functions of

$$r = \left[\sum_{i=1}^{4} (x^i)^2 \right]^{1/2},$$

where the coordinates $x^1, \dots, x^4$ are such that they become Cartesian coordinates in the limit of vanishing curvature. More specifically, we describe the flat metric by the oriented orthonormal frame dr, $-\frac{1}{2}r\tilde{\zeta}^1$, $-\frac{1}{2}r\tilde{\zeta}^2$, $-\frac{1}{2}r\tilde{\zeta}^3$ with $\tilde{\zeta}^j$ given by (10.32) and write

$$(e^1, \dots, e^4) = (f(r)\,dr, \, -\tfrac{1}{2}r\tilde{\zeta}^1, \, -\tfrac{1}{2}r\tilde{\zeta}^2, \, -\tfrac{1}{2}rg(r)\tilde{\zeta}^3). \tag{10.47}$$

This ansatz leads to the following expression for the spin connection:

$$\omega^{21} = \frac{1}{rf}e^2, \qquad\qquad \omega^{34} = \frac{g}{r}e^2,$$

$$\omega^{31} = \frac{1}{rf}e^3, \qquad\qquad \omega^{42} = \frac{g}{r}e^3, \tag{10.48}$$

$$\omega^{41} = \left(\frac{1}{rf} + \frac{dg/dr}{fg} \right)e^4, \quad \omega^{23} = \frac{2-g^2}{rg}e^4.$$

Imposing the condition $\tilde{\omega} = -\omega$ (requiring $\tilde{\omega} = \omega$ would reverse the orientation of our frame) we find the equations

$$fg = 1,$$
$$g + r\frac{dg}{dr} = f(2 - g^2). \tag{10.49}$$

The solution reads

$$g(r) = f^{-1}(r) = [1 - (a/r)^4]^{1/2}, \quad r > a, \tag{10.50}$$

with a constant of integration a. For $r \to \infty$, g and f tend to 1; i.e. the metric is asymptotically flat. This is analogous to the Yang–Mills case where the instanton potential becomes a pure gauge as $r \to \infty$. The singularity at $r = a$ can be understood as a coordinate singularity by considering $x^1, \dots, x^4$ as coordinates on a suitable manifold, namely the cotangent bundle of S^2. For a detailed discussion of this point and further properties of the solution we must refer the reader to Eguchi & Hanson (1979).

10.5 The Dirac monopole as a connection on a nontrivial principal bundle

We have seen in section 10.1 that the local potentials A_1, A_2 (cf. (10.10),

(10.11)) describing the Dirac monopole may be considered as representing a connection on a $U(1)$ principal bundle over S^2 characterized by the transition function

$$g_{21}(x) = e^{im\varphi}, \quad x \in \mathcal{U}_1 \cap \mathcal{U}_2, \quad m \in \mathbb{Z}. \qquad (10.51)$$

(We continue to use the notation of section 10.1.) In section 9.5 we have already discussed the $U(1)$ bundle over S^2 with transition function $g_{21}(x) = e^{i\varphi}$. So we restrict ourselves to the case $m = 1$ and look for a connection 1-form $\mathscr{A}$ on the total space of this bundle, which is S^3, such that

$$\sigma_1^* \mathscr{A} = A_1 = \tfrac{1}{2}i(1 - \cos \vartheta)\,d\varphi,$$
$$\sigma_2^* \mathscr{A} = A_2 = \tfrac{1}{2}i(-1 - \cos \vartheta)\,d\varphi. \qquad (10.52)$$

Here, σ_r is the local section associated with the trivialization over $\mathcal{U}_r$ (cf. (9.39)). In the notation of section 9.5 they read:

$$\sigma_1(\sin \vartheta \cos \varphi, \sin \vartheta \sin \varphi, \cos \vartheta) = (\cos \tfrac{1}{2}\vartheta, e^{i\varphi} \sin \tfrac{1}{2}\vartheta), \quad \vartheta < \pi,$$
$$\qquad (10.53)$$
$$\sigma_2(\sin \vartheta \cos \varphi, \sin \vartheta \sin \varphi, \cos \vartheta) = (e^{-i\varphi} \cos \tfrac{1}{2}\vartheta, \sin \tfrac{1}{2}\vartheta), \quad 0 < \vartheta.$$

In the following we shall need a convenient description of the Lie algebra $u(1)$. Choosing a coordinate α on $U(1) - \{-1\}$ given by

$$U(1) - \{-1\} \to (-\pi, \pi)$$
$$\qquad (10.54)$$
$$e^{i\alpha} \mapsto \alpha,$$

one easily confirms that each left invariant vector field B on $U(1)$, i.e. each Lie algebra element, may be represented as

$$B(z) = b \frac{\partial}{\partial \alpha}(z), \quad z \in U(1) - \{-1\}, \qquad (10.55)$$

with some $b \in \mathbb{R}$. According to our convention of using antihermitian generators for unitary groups, we identify $B \in u(1)$ with the imaginary number ib.

For $B \in u(1)$ the fundamental vector field $\lambda B \in \text{vect}(S^3)$ is defined by (cf. (9.54))

$$(\lambda B)(z_1, z_2) = T_1 \, \text{bit}_{(z_1, z_2)}(B(1)) \qquad (10.56)$$

where

$$\text{bit}_{(z_1, z_2)}(e^{i\alpha}) = \tilde{R}_{e^{i\alpha}}(z_1, z_2) = (z_1 e^{i\alpha}, z_2 e^{i\alpha}), \quad (z_1, z_2) \in S^3, \qquad (10.57)$$

(cf. (9.43)). We can imagine $\tilde{R}_{e^{i\alpha}}$ as being defined on all $\mathbb{C}^2$. Correspondingly, we consider λB as a vector field on $\mathbb{C}^2$ which on S^3 is tangent to this sphere, because $\mathrm{bit}_{(z_1, z_2)}$ with $(z_1, z_2) \in S^3$ maps $U(1)$ into S^3. Introducing real coordinates $y^1, \ldots, y^4$ on $\mathbb{C}^2$ by putting

$$z_1 = y^1 + iy^2, \quad z_2 = y^3 + iy^4 \tag{10.58}$$

we find

$$(\lambda B)(z_1, z_2) = b\left(-y^2 \frac{\partial}{\partial y^1} + y^1 \frac{\partial}{\partial y^2} - y^4 \frac{\partial}{\partial y^3} + y^3 \frac{\partial}{\partial y^4} \right) \tag{10.59}$$

with $b \in \mathbb{R}$ as in (10.55).

We define

$$\mathscr{A} := i(y^1 \, dy^2 - y^2 \, dy^1 + y^3 \, dy^4 - y^4 \, dy^3) \tag{10.60}$$

and claim that this 1-form when considered on S^3 is the connection 1-form we are looking for (Trautman 1977). Indeed, $\mathscr{A}$ takes imaginary values, which we identify with elements of $u(1)$ as explained above. Furthermore, one verifies immediately

$$\mathscr{A}_{(z_1, z_2)}((\lambda B)(z_1, z_2)) = ib, \tag{10.61}$$

property (b) in the theorem of section 9.8. Property (c) reads in our special case

$$\tilde{R}^*_{e^{i\alpha}} \mathscr{A} = \mathrm{Ad}(e^{-i\alpha}) \mathscr{A} = \mathscr{A} \tag{10.62}$$

and is easily confirmed. The smoothness of $\mathscr{A}$ being obvious we conclude that $\mathscr{A}$ describes a connection on our bundle. By explicit calculation one verifies (10.52) proving our claim. So the potentials A_1 and A_2 of the Dirac monopole with $m = 1$ are local representatives of the connection (10.60).

10.6 Recovering the Dirac monopole from the 't Hooft–Polyakov monopole

We now attack the problem mentioned at the end of section 10.2: How can we discover the nontrivial $U(1)$ bundle of the Dirac monopole with $m = 2$ inside the trivial $SO(3)$ bundle of the 't Hooft–Polyakov monopole? As an example where a similar 'embedding' occurs consider a Möbius strip, a nontrivial bundle over S^1 with standard fibre $[-1, 1]$, lying in a trivial bundle over S^1 whose standard fibre is the closed disk $B^2 = \{(x^1, x^2) \in \mathbb{R}^2 \mid (x^1)^2 + (x^2)^2 \leqslant 1\}$.

The base space of the Dirac monopole bundle constructed above is S^2. Therefore it will be advantageous to restrict the fields of the 't Hooft–Polyakov ansatz (10.17) to a 2-sphere of some radius r in 3-space. Since A

vanishes on radial 3-vectors this is a natural thing to do. Henceforth we shall identify this sphere with the unit sphere S^2, which will serve as the base space of our bundles.

The ansatz for the scalar fields is invariant under gauge transformations of the form

$$\gamma(x) = \exp\left(-\beta(x) \sum_{j=1}^{3} x^j T_j \right), \quad x \in S^2, \tag{10.63}$$

with $\beta: S^2 \to \mathbb{R}$ and T_j as in (10.13), because $\gamma(x)$ represents a rotation by an angle $\beta(x)$ around the radial direction, i.e. the direction of Φ. This is the $U(1)$ gauge invariance of electromagnetism embedded in the symmetry group $SO(3)$ of the Lagrangian (10.14) via the family of homomorphisms

$$U(1) \to SO(3)$$

$$e^{i\alpha} \mapsto \exp\left(-\alpha \sum_{j=1}^{3} x^j T_j \right). \tag{10.64}$$

We now look for a $U(1)$ principal bundle over S^2 such that the gauge transformations in this bundle correspond to the unbroken $U(1)$ gauge transformations (10.63) of the 't Hooft–Polyakov monopole.

On one side we have a trivial $SO(3)$ principal bundle over S^2:

$$\tilde{P} = S^2 \times SO(3). \tag{10.65}$$

A gauge transformation is given by

$$S^2 \times SO(3) \to S^2 \times SO(3)$$

$$(x, W) \mapsto (x, \gamma(x) W) \tag{10.66}$$

with $\gamma: S^2 \to SO(3)$ (cf. (9.168)). In particular, we can choose γ as in (10.63).

On the other side, we look for a (possibly nontrivial) $U(1)$ principal bundle (P, S^2, π). Since the open subsets $\mathcal{U}_1 = S^2 - \{(0, 0, -1)\}$, $\mathcal{U}_2 = S^2 - \{(0, 0, 1)\}$ are contractible, the portion $\pi^{-1}(\mathcal{U}_r)$ of P is trivial (cf. section 9.1), and we can use $\mathcal{U}_1$, $\mathcal{U}_2$ as trivializing neighbourhoods. A gauge transformation in P is then given by maps $\gamma_r: \mathcal{U}_r \to U(1)$ satisfying (9.155). Since $U(1)$ is abelian, this means

$$\gamma_1(x) = \gamma_2(x), \quad x \in \mathcal{U}_1 \cap \mathcal{U}_2; \tag{10.67}$$

i.e. we have one funcion $\tilde{\gamma}: S^2 \to U(1)$, from which we can construct the map $f: P \to P$ introduced in section 9.14 according to the formula

$$f(p) = p\tilde{\gamma}(\pi(p)). \tag{10.68}$$

To connect both sides we require the existence of a bundle map $(f_P, \varphi, \mathrm{id}_{S^2})$,

$$f_P: P \to \tilde{P},$$

$$\varphi: U(1) \to SO(3),$$

(10.69)

such that under f_P the gauge transformation (10.68) with $\tilde{\gamma}(x) = e^{i\beta(x)}$, $\beta: S^2 \to \mathbb{R}$, corresponds to the gauge transformation in $\tilde{P}$ given by (10.66) with γ as in (10.63):

$$S^2 \times SO(3) \to S^2 \times SO(3)$$

(10.70)

$$(x, W) \mapsto \left(x, \exp\left(-\beta(x) \sum_{j=1}^{3} x^j T_j \right) W \right).$$

The properties (b), (c) in the definition of a bundle map of principal bundles (cf. section 9.4) enable us to write

$$f_P(p) = (\pi(p), G(p)), \quad p \in P,$$

(10.71)

where $G: P \to SO(3)$ satisfies

$$G(pe^{i\alpha}) = G(p)\varphi(e^{i\alpha}).$$

(10.72)

Starting from $p \in \pi^{-1}(x) \subset P$, applying first the gauge transformation f and then the map f_P we arrive at

$$(x, G(p)\varphi(e^{i\beta(x)})) \in \tilde{P}.$$

(10.73)

On the other hand, acting with the gauge transformation (10.70) on $f_P(p)$ we get

$$\left(x, \exp\left(-\beta(x) \sum_{j=1}^{3} x^j T_j \right) G(p) \right) \in \tilde{P}.$$

(10.74)

These two points in $\tilde{P}$ are required to coincide (for all $p \in P$).

To proceed further, we introduce local trivializations of the still unknown bundle P:

$$\pi^{-1}(\mathcal{U}_r) \to \mathcal{U}_r \times U(1)$$

$$p \mapsto (\pi(p), f_r(p)).$$

(10.75)

So, for $p \in \pi^{-1}(\mathcal{U}_r)$ we can write

$$G(p) = \tilde{G}_r(\pi(p), f_r(p)).$$

(10.76)

In terms of $\tilde{G}_r$, (10.72) reads

$$\tilde{G}_r(x, e^{i\alpha' + i\alpha}) = \tilde{G}_r(x, e^{i\alpha'})\varphi(e^{i\alpha}), \tag{10.77}$$

and equality of (10.73) and (10.74) means

$$\tilde{G}_r(x, e^{i\alpha'})\varphi(e^{i\beta}) = \exp\left(-\beta \sum_{j=1}^{3} x^j T_j\right)\tilde{G}_r(x, e^{i\alpha'}). \tag{10.78}$$

Equation (10.78) can be fulfilled only if

$$\varphi(e^{i\beta}) = \exp\left(-\beta \sum_{j=1}^{3} n^j T_j\right) \tag{10.79}$$

with $(n^1)^2 + (n^2)^2 + (n^3)^2 = 1$. We choose $n^3 = 1$ and find as a solution to (10.78)

$$\tilde{G}_1(x, e^{i\alpha}) = R\left(\begin{pmatrix} \cos\frac{1}{2}\vartheta & -e^{-i\varphi}\sin\frac{1}{2}\vartheta \\ e^{i\varphi}\sin\frac{1}{2}\vartheta & \cos\frac{1}{2}\vartheta \end{pmatrix}\right)\exp(-\alpha T_3), \quad \vartheta < \pi,$$

$$\tilde{G}_2(x, e^{i\alpha}) = R\left(\begin{pmatrix} e^{-i\varphi}\cos\frac{1}{2}\vartheta & -\sin\frac{1}{2}\vartheta \\ \sin\frac{1}{2}\vartheta & e^{i\varphi}\cos\frac{1}{2}\vartheta \end{pmatrix}\right)\exp(-\alpha T_3), \quad 0 < \vartheta, \tag{10.80}$$

where ϑ, φ are polar coordinates and $R: SU(2) \to SO(3)$ is the same homomorphism as in section 10.2. For $p \in \pi^{-1}(\mathcal{U}_1) \cap \pi^{-1}(\mathcal{U}_2)$ we must have

$$G(p) = \tilde{G}_1(\pi(p), f_1(p))$$

$$= \tilde{G}_2(\pi(p), f_2(p)). \tag{10.81}$$

This relation determines the transition function of P. One finds

$$f_2(p) = e^{2i\varphi} f_1(p) \tag{10.82}$$

or

$$g_{21}(x) = e^{2i\varphi}, \quad x \in \mathcal{U}_1 \cap \mathcal{U}_2, \tag{10.83}$$

where φ is the azimuthal angle belonging to $x = \pi(p)$. The bundle P carrying the unbroken $U(1)$ gauge invariance of the 't Hooft–Polyakov monopole has turned out to be identical to the bundle of the Dirac monopole with $m = 2$. Equation (10.80) shows that $U(1)$ is mapped into $SO(3)$ in a twisted manner, i.e. in a way which depends on the base point x.

So much for the bundles. What happens to the connections under the map f_P? On P we have a connection $\mathscr{A}$, which is represented locally by the potentials (10.10), (10.11) for $m = 2$. One can show that there is exactly one connection $\tilde{\mathscr{A}}$ on $\tilde{P}$ with

$$f_P^* \tilde{\mathscr{A}} = (T_1\varphi)\mathscr{A}. \tag{10.84}$$

Here, $T_1\varphi: T_1U(1) \to T_eSO(3)$ is interpreted as a Lie algebra homomorphism mapping $ib \in u(1)$ onto $-bT_3 \in so(3)$. What does $\tilde{\mathscr{A}}$ look like? Applying the pull back by the trivializing section σ_r to (10.84), we get

$$(f_P \circ \sigma_r)^* \tilde{\mathscr{A}} = (T_1\varphi)A_r \qquad (10.85)$$

where for $x \in \mathscr{U}_r$

$$(f_P \circ \sigma_r)(x) = (x, \tilde{G}_r(x, 1)). \qquad (10.86)$$

Since $\tilde{\mathscr{A}}$ is a connecion on a trivial bundle, we have (cf. (9.76))

$$\tilde{\mathscr{A}}_{(x,W)} = W^{-1}\tilde{A}_x W + W^{-1}\,dW, \quad (x, W) \in S^2 \times SO(3); \qquad (10.87)$$

with an $so(3)$-valued 1-form $\tilde{A}$ on S^2. Consequently, (10.85) leads to

$$(T_1\varphi)A_{r,x} = \tilde{G}_r(x, 1)^{-1}\tilde{A}_x\tilde{G}_r(x, 1) + \tilde{G}_r(x, 1)^{-1}\,d\tilde{G}_r(x, 1), \qquad (10.88)$$

i.e. we get $(T_1\varphi)A_r$ from $\tilde{A}$ by a gauge transformation which is well defined only on $\mathscr{U}_r$. Actually, $\tilde{G}_1(x, 1)^{-1}$ coincides with $\gamma(x)$ in (10.24), and one verifies by straightforward calculation (cf. (10.25)) that $\tilde{A}$ is the potential of the 't Hooft–Polyakov ansatz on the sphere at infinity. So we recover within the bundle formalism the result that the 't Hooft–Polyakov monopole becomes a Dirac monopole with $m = 2$ as $r \to \infty$.

What changes if we take $SU(2)$ instead of $SO(3)$ in the above considerations? Equation (10.79) with T_j replaced by $\tau_j = -\frac{1}{2}i\sigma_j$ (cf. (8.55)) does not define a group homomorphism. So we have to modify the embedding (10.64) of $U(1)$ by writing

$$U(1) \to SU(2)$$

$$e^{i\alpha} \mapsto \exp\left(-2\alpha \sum_{j=1}^{3} x^j \tau_j\right). \qquad (10.89)$$

The final consequence is that we get the bundle of the Dirac monopole with $m = 1$. In physical terms: The electric charge of an $SU(2)$ doublet is $\frac{1}{2}g$ whereas the electric charge of a triplet is g.

10.7 The instanton bundle

In section 10.3 we derived the instanton potential

$$A_1 = \frac{r^2}{r^2 + c^2}\gamma^{-1}\,d\gamma \qquad (10.90)$$

as a solution to the Euclidean Yang–Mills equation. The map $\gamma: \mathbb{R}^4 - \{0\} \to SU(2)$ is given by (10.30), and r is the distance from the origin (cf. (10.28)). The potential (10.90) is regular at $x = 0$ but decays only as r^{-1} for $r \to \infty$. By a gauge transformation with the map γ we can change these features. The resulting potential

$$A_2 = \gamma A_1 \gamma^{-1} + \gamma \, \mathrm{d}\gamma^{-1} = \frac{c^2}{r^2 + c^2} \gamma \, \mathrm{d}\gamma^{-1} \qquad (10.91)$$

is singular at $x = 0$ but vanishes more rapidly (like r^{-3}) at infinity.

Identifying $\mathbb{R}^4$ with $S^4 - \{\text{south pole}\} = \mathscr{U}_1$ by means of the stereographic projection α_1 (equation (7.9) with $n = 4$), we can consider A_1 as being defined on $\mathscr{U}_1$ (via pullback by α_1). The slow decay of A_1 as $r \to \infty$ prevents us from extending A_1 smoothly to all of S^4. Similarly, we may imagine A_2 as living on $\mathscr{U}_2 = S^4 - \{\text{north pole}\}$: A_2 vanishes sufficiently fast at infinity to become regular at the south pole, but the singularity at the origin manifests itself as a singularity at the north pole. On $\mathscr{U}_1 \cap \mathscr{U}_2$ the potentials A_1 and A_2 are connected by a gauge transformation. This situation is very much like in the case of the Dirac monopole. Again, the theorem in section 9.10 tells us that A_1 and A_2 are the local representatives of a connection on an $SU(2)$ principal bundle over S^4 whose transition function g_{21} is γ. The total space of this bundle is S^7 (Trautman 1977).

We have considered only the instanton solution. However, it has been proved (Uhlenbeck 1979) that any finite action solution of the Euclidean Yang–Mills equation leads to a fibre bundle over S^4.

10.8 Chern classes

An important means for classifying fibre bundles are the so-called characteristic classes, in particular the Chern classes, which we shall now discuss.

Let G be a Lie group, $\mathfrak{g}$ its Lie algebra. A symmetric j-linear mapping

$$I : \underbrace{\mathfrak{g} \times \mathfrak{g} \times \cdots \times \mathfrak{g}}_{j \text{ factors}} \to \mathbb{R} \text{ or } \mathbb{C} \qquad (10.92)$$

is called G-invariant or simply invariant, if for all $g \in G$ and $A_1, A_2, \ldots, A_j \in \mathfrak{g}$

$$I(\mathrm{Ad}_g A_1, \mathrm{Ad}_g A_2, \ldots, \mathrm{Ad}_g A_j) = I(A_1, A_2, \ldots, A_j). \qquad (10.93)$$

By $\mathscr{I}^j(G)$ we denote the set of all G-invariant symmetric j-linear forms (10.92). We introduce the following abbreviation:

$$I(A^j) := I(A, A, \ldots, A), \qquad (10.94)$$

where $I \in \mathscr{I}^j(G)$. The expression $I(A^j)$ is called a G-invariant polynomial on $\mathfrak{g}$.

With the help of a basis T_a of g we can write

$$A_i = \sum_a A_i^a T_a, \quad A_i^a \in \mathbb{R},$$

and get

$$I(A_1, A_2, \ldots, A_j) = \sum_{a_1, \ldots, a_j} A_1^{a_1} \cdots A_j^{a_j} I(T_{a_1}, \ldots, T_{a_j}), \qquad (10.95)$$

in particular,

$$I(A^j) = \sum_{a_1, \ldots, a_j} A^{a_1} \cdots A^{a_j} I(T_{a_1}, \ldots, T_{a_j}). \qquad (10.96)$$

So $I(A^j)$ is a homogeneous polynomial of degree j in the components of A with respect to a basis of g. As (10.95) and (10.96) show, I is completely determined, once $I(A^j)$ is known for all $A \in g$: $I(A_1, A_2, \ldots, A_j)$ is the complete polarization of $I(A^j)$.

Now let M be an n-dimensional manifold and $\varphi_i \in \Lambda^{q_i}(M, g)$, $i = 1, 2, \ldots, j$. Writing

$$\varphi_i = \sum_a \varphi_i^a T_a, \quad \varphi_i^a \in \Lambda^{q_i} M,$$

we define

$$I(\varphi_1, \varphi_2, \ldots, \varphi_j) := \sum_{a_1, \ldots, a_j} \varphi_1^{a_1} \wedge \cdots \wedge \varphi_j^{a_j} I(T_{a_1}, \ldots, T_{a_j}), \qquad (10.97)$$

a $(q_1 + q_2 + \cdots + q_j)$-form on M. If G is a matrix group we can consider φ_i as a matrix of q_i-forms. Then we get $I(\varphi_1, \varphi_2, \ldots, \varphi_j)$ by inserting these matrices in I and interpreting the products as exterior products.

Let (P, M, π) be a principal bundle with structure group G and $\mathscr{A}$ a connection on P. On a trivializing neighbourhood we can describe $\mathscr{A}$ by a g-valued 1-form A and the corresponding curvature by $F = dA + \frac{1}{2}[A, A]$. For $I \in \mathscr{I}^j(G)$, the 2j-form $I(F^j)$ does not depend on the particular trivialization due to (9.102) and the invariance of I. Consequently, the various locally defined forms $I(F^j)$ fit together to yield a 2j-form on all of M, which we again denote by $I(F^j)$. One can show that due to the Bianchi identity this form is closed:

$$dI(F^j) = 0. \qquad (10.98)$$

Hence, $I(F^j)$ determines a cohomology class in $H^{2j}(M)$ (or its complex analogue $H^{2j}(M, \mathbb{C})$).

If $\mathscr{A}$ is another connection on P with a local potential $\mathring{A}$ and curvature $\mathring{F}$, we define an interpolation between $\mathring{A}$ and A by

$$A_\tau := \mathring{A} + \tau(A - \mathring{A}), \quad 0 \leqslant \tau \leqslant 1, \qquad (10.99)$$

and a corresponding field strength

$$F_\tau = dA_\tau + \frac{1}{2}[A_\tau, A_\tau]. \qquad (10.100)$$

Then we have the Chern–Simons formula

$$I(F^j) - I(\mathring{F}^j) = dQ(A, \mathring{A}) \tag{10.101}$$

with

$$Q(A, \mathring{A}) := j \int_0^1 d\tau\, I(A - \mathring{A}, F_\tau, \ldots, F_\tau). \tag{10.102}$$

Since $A - \mathring{A}$ behaves like F under gauge transformations, $I(A - \mathring{A}, F_\tau, \ldots, F_\tau)$ and consequently also $Q(A, \mathring{A})$ define a $(2j-1)$-form on all of M. Equation (10.101) shows that the cohomology class of $I(F^j)$ does not depend on the connection: The difference of $I(F^j)$ and $I(\mathring{F}^j)$ is an exact form. The class of $I(F^j)$ characterizes the bundle P and is called a characteristic class. We remark in passing that $Q(A, \mathring{A})$ gives rise to a secondary (or Chern–Simons) characteristic class.

Consider now the special case where the structure group G of the bundle P is $GL(m, \mathbb{C})$, the group of all nonsingular $m \times m$ matrices with complex coefficients (or a closed subgroup of $GL(m, \mathbb{C})$). The elements of g are identified with complex $m \times m$ matrices. We define invariant polynomials $c_j \in \mathscr{I}^j(GL(m, \mathbb{C}))$, $j = 1, 2, \ldots, m$, by

$$\det\left(\lambda \mathbb{1} - \frac{1}{2\pi i} B\right) = \sum_{j=0}^m c_j(B^j)\lambda^{m-j}, \quad B \in gl(m, \mathbb{C}). \tag{10.103}$$

If the matrices in G are unitary, the Lie algebra elements B are antihermitian and $c_j(B^j)$ turns out to be real. The cohomology class of $c_j(F^j)$, an element of $H^{2j}(M)$, does not depend on the chosen connection, as shown above, and is called the jth Chern class of P. One can prove that any characteristic class of P may be expressed as a polynomial in the Chern classes. Explicitly, we have

$$c_j(F^j) = \frac{(-1)^j}{(2\pi i)^j j!} \sum_{i_1,\ldots,i_j=1}^m \sum_{\pi \in \mathscr{S}_j} \text{sig}\,\pi F^{i_1}{}_{i_{\pi(1)}} \wedge \cdots \wedge F^{i_j}{}_{i_{\pi(j)}}, \tag{10.104}$$

in particular,

$$c_1 = -\frac{1}{2\pi i} \text{tr}\, F, \tag{10.105}$$

$$c_2 = \frac{1}{8\pi^2}(\text{tr}(F \wedge F) - (\text{tr}\, F) \wedge (\text{tr}\, F)). \tag{10.106}$$

Note that for $G = SU(m)$

$$c_1 = 0, \quad c_2 = \frac{1}{8\pi^2} \text{tr}(F \wedge F).$$

The Chern–Simons formula (10.101) applied to $I(F^2) = \text{tr}(F \wedge F)$ gives the result

$$\operatorname{tr}(F \wedge F) - \operatorname{tr}(\overset{\circ}{F} \wedge \overset{\circ}{F})$$
$$= \operatorname{d}\operatorname{tr}(2\alpha \wedge \overset{\circ}{F} + \alpha \wedge \operatorname{d}\alpha + 2\alpha \wedge \overset{\circ}{A} \wedge \alpha + \tfrac{2}{3}\alpha \wedge \alpha \wedge \alpha) \qquad (10.107)$$

where $\alpha = A - \overset{\circ}{A}$. For $\overset{\circ}{A} = 0$ we find

$$\operatorname{tr}(F \wedge F) = \operatorname{d}\operatorname{tr}((\operatorname{d}A) \wedge A + \tfrac{2}{3}A \wedge A \wedge A), \qquad (10.108)$$

an equation which will be useful later on. It expresses the well-known fact that the axial $U(1)$-anomaly can be written as a total divergence.

The coefficients in the definition of the c_j are chosen such that integrals of c_j over compact, $2j$-dimensional submanifolds of M without boundary are integer. Obviously, these integrals depend only on the cohomology class of c_j. If M is compact without boundary and $c \in \Lambda^n M$, a polynomial in c_j, represents a characteristic class, we call $\int_M c$ a Chern number. For $n = 4$ we have only two independent Chern numbers:

$$\int_M c_2, \quad \int_M c_1 \wedge c_1.$$

Consider, for instance, the $U(1)$ bundle over S^2 determined by the local potentials (10.10), (10.11) of the Dirac monopole. We get for the only nontrivial Chern number (cf. (2.113)):

$$\int_{S^2} c_1 = -\frac{1}{2\pi i} \int_{S^2} F = -\frac{m}{4\pi} \int_{S^2} (x^1 \operatorname{d}x^2 \wedge \operatorname{d}x^3 + x^2 \operatorname{d}x^3 \wedge \operatorname{d}x^1$$
$$+ x^3 \operatorname{d}x^1 \wedge \operatorname{d}x^2) = -m. \qquad (10.109)$$

More instructively, we can arrive at the same result in the following way. Denoting by $S^n_+ (S^n_-)$ the upper (lower) hemisphere of S^n,

$$S^n_+ := \{(x^1, x^2, \ldots, x^{n+1}) \in S^n \mid x^{n+1} \geqslant 0\},$$
$$S^n_- := \{(x^1, x^2, \ldots, x^{n+1}) \in S^n \mid x^{n+1} \leqslant 0\}, \qquad (10.110)$$

we may write

$$\int_{S^2} c_1 = -\frac{1}{2\pi i} \int_{S^2_+} F - \frac{1}{2\pi i} \int_{S^2_-} F. \qquad (10.111)$$

On $S^2_+ (S^2_-)$ the potential $A_1 (A_2)$ is well defined. So we obtain, applying Stokes' theorem and paying due attention to the orientations:

$$\int_{S^2} c_1 = -\frac{1}{2\pi i} \int_{S^2_+} \operatorname{d}A_1 - \frac{1}{2\pi i} \int_{S^2_-} \operatorname{d}A_2 = -\frac{1}{2\pi i} \int_{S^1} A_1 + \frac{1}{2\pi i} \int_{S^1} A_2$$

$$= -\frac{1}{2\pi i} \int_{S^1} \gamma^{-1} \operatorname{d}\gamma = -m \qquad (10.112)$$

with γ given by (10.8). Only the transition function γ enters this calculation, as it must be, since a Chern number is a characteristic property of the bundle, not of the connection. Potentials with different values of the 'topological quantum number' m stem from connections on different principal bundles.

We now perform a computation analogous to (10.111), (10.112) for the instanton bundle from section 10.7. Since $G = SU(2)$, the only nontrivial Chern class is represented by

$$c_2 = \frac{1}{8\pi^2} \operatorname{tr}(F \wedge F).$$

The corresponding Chern number, which up to the sign agrees with the so-called topological charge, can be written as

$$\int_{S^4} c_2 = \frac{1}{8\pi^2} \int_{S_+^4} \operatorname{tr}(F_1 \wedge F_1) + \frac{1}{8\pi^2} \int_{S_-^4} \operatorname{tr}(F_2 \wedge F_2). \qquad (10.113)$$

Here we may set $F_r = dA_r + A_r \wedge A_r$, $r = 1, 2$, with A_1 and A_2 as in section 10.7. Using (10.108) for F_1 on S_+^4 and for F_2 on S_-^4, we get

$$\int_{S^4} c_2 = \frac{1}{8\pi^2} \int_{S^3} \operatorname{tr}((dA_1) \wedge A_1 + \tfrac{2}{3} A_1 \wedge A_1 \wedge A_1)$$

$$- \frac{1}{8\pi^2} \int_{S^3} \operatorname{tr}((dA_2) \wedge A_2 + \tfrac{2}{3} A_2 \wedge A_2 \wedge A_2)$$

$$= -\frac{1}{24\pi^2} \int_{S^3} \operatorname{tr}(\gamma^{-1}(d\gamma) \wedge \gamma^{-1}(d\gamma) \wedge \gamma^{-1}(d\gamma)) \qquad (10.114)$$

with γ given by (10.30). This expression counts how often the image of S^3 under γ covers $SU(2)$; it is the winding number of γ. As above, only the transition function γ of the underlying bundle appears in the final formula. With the help of (10.31) we can evaluate it further:

$$\int_{S^4} c_2 = -\frac{1}{2\pi^2} \int_{S^3} (-\tfrac{1}{2}\tilde{\zeta}^1) \wedge (-\tfrac{1}{2}\tilde{\zeta}^2) \wedge (-\tfrac{1}{2}\tilde{\zeta}^3). \qquad (10.115)$$

Since $-\tfrac{1}{2}\tilde{\zeta}^1$, $-\tfrac{1}{2}\tilde{\zeta}^2$, $-\tfrac{1}{2}\tilde{\zeta}^3$ (with $r = 1$) form an oriented, orthonormal frame on S^3 with respect to the standard metric, the 3-form $(-\tfrac{1}{2}\tilde{\zeta}^1) \wedge (-\tfrac{1}{2}\tilde{\zeta}^2) \wedge (-\tfrac{1}{2}\tilde{\zeta}^3)$ is the volume element for S^3, and we find

$$\int_{S^4} c_2 = -1. \qquad (10.116)$$

Finally, we remark that an $SU(2)$ bundle over S^4 with transition function

$g_{21} = \gamma^m$, $m \in \mathbb{Z}$, has

$$\int_{S^4} c_2 = -m. \qquad (10.117)$$

This follows inductively from the identity

$$\operatorname{tr}((\gamma_1 \gamma_2)^{-1} d(\gamma_1 \gamma_2) \wedge (\gamma_1 \gamma_2)^{-1} d(\gamma_1 \gamma_2) \wedge (\gamma_1 \gamma_2)^{-1} d(\gamma_1 \gamma_2))$$

$$= \sum_{i=1}^{2} \operatorname{tr}(\gamma_i^{-1}(d\gamma_i) \wedge \gamma_i^{-1}(d\gamma_i) \wedge \gamma_i^{-1}(d\gamma_i))$$

$$+ d\{3 \operatorname{tr}(\gamma_1^{-1}(d\gamma_1) \wedge \gamma_2(d\gamma_2^{-1}))\}. \qquad (10.118)$$

Problems

10.1 Apply the gauge transformation (10.24) to the 't Hooft–Polyakov ansatz (10.17).

10.2 Verify (10.31).

10.3 Calculate the field strength belonging to the instanton ansatz (10.29).

10.4 Show that (10.48) is the Riemannian connection belonging to the metric described by the ansatz (10.47).

10.5 Verify the expression (10.59) for the fundamental vector field λB.

10.6 Repeat the calculations of section 10.6 with the symmetry group $SU(2)$ instead of $SO(3)$.

10.7 Prove (10.98) in the special case where G is a matrix group and $I(F^2) = \operatorname{tr}(F \wedge F)$.

10.8 Derive (10.107) from the Chern–Simons formula.

10.9 Verify (10.114).

11

Spin

In chapter 8 we constructed the universal covering group of the rotation group in three dimensions. In special relativity we need the universal covering group of the Lorentz group to describe spin $\frac{1}{2}$. For convenience we discuss immediately the case of arbitrary dimension and signature of the metric. We then define spinors and the Dirac operator in the setting of general relativity. As always we start with linear algebra, then consider open subsets of $\mathbb{R}^n$, which in a third step are glued together.

11.1 Clifford algebras

Let V be a real vector space of dimension n equipped with a metric g which has r plus signs and s minus signs, $r + s = n$. We choose an orthonormal basis e^i, $i = 1, 2, \ldots, n$:

$$g(e^i, e^j) = \eta^{ij}, \tag{11.1}$$

$$\eta = \left(\begin{array}{ccc|ccc} 1 & & & & & \\ & \ddots & & & & \\ & & 1 & & & \\ \hline & & & -1 & & \\ & & & & \ddots & \\ & & & & & -1 \end{array} \right) \begin{array}{c} \\ \\ \Big\} r \\ \\ \\ \Big\} s \end{array} . \tag{11.2}$$

We use superscripts because later V will be a cotangent space. The Clifford algebra $C(r, s)$ is by definition the real associative algebra with neutral element $\mathbb{1}$ generated by the elements of V with the relation

$$vw + wv = 2g(v, w)\mathbb{1}, \tag{11.3}$$

where the Clifford product is denoted by juxtaposition. The neutral element on the rhs is often omitted. Note also that many authors use the opposite sign on the rhs of (11.3). As a vector space, the Clifford algebra has the following direct sum decomposition:

$$C(r, s) = \bigoplus_{\alpha = 0}^{n} C^\alpha. \tag{11.4}$$

C^0 is spanned by $\mathbb{1}$ and identified with $\mathbb{R}$. The subspace C^α for $\alpha > 0$ has a basis $e^{i_1} e^{i_2} \cdots e^{i_\alpha}$, $i_1 < i_2 < \cdots < i_\alpha$. In particular, $C^1 = V$,

$$\dim C^\alpha = \binom{n}{\alpha}, \tag{11.5}$$

$$\dim C(r, s) = 2^n. \tag{11.6}$$

We define the even subalgebra C_e and the odd subspace C_o by

$$C_e := \bigoplus_{\alpha \text{ even}} C^\alpha, \tag{11.7}$$

$$C_o := \bigoplus_{\alpha \text{ odd}} C^\alpha \tag{11.8}$$

yielding a $\mathbb{Z}_2$-grading of $C(r, s)$. Some authors allow also degenerate bilinear forms g on V, in which case a Grassmann algebra is a particular kind of Clifford algebra with $g = 0$.

Consider $C(0, 2)$. A basis of C^0, C^1, and C^2 is given by

$$\mathbb{1},$$

$$e^1 =: \hat{i}, \quad e^2 =: \hat{j},$$

and

$$e^1 e^2 =: \hat{k},$$

respectively, with multiplication rules

$$\hat{i}\hat{j} = \hat{k}, \tag{11.9}$$

$$\hat{k}\hat{i} = e^1 e^2 e^1 = e^2 = \hat{j}, \tag{11.10}$$

$$\hat{j}\hat{k} = e^2 e^1 e^2 = e^1 = \hat{i}. \tag{11.11}$$

Therefore $C(0, 2)$ is isomorphic to the algebra of quaternions. Note that $C(0, 2)$ is not isomorphic to $C(2, 0)$. However, their complexifications (replacing $\mathbb{R}$ by $\mathbb{C}$) are both isomorphic to $\mathbb{C}(2)$, the associative algebra of 2×2 matrices with complex coefficients.

Now consider $C(1, 1)$. It is isomorphic to $\mathbb{R}(2)$, the real algebra of 2×2 matrices,

$$C(1, 1) \stackrel{\cong}{\to} \mathbb{R}(2),$$

$$\mathbb{1} \mapsto \begin{pmatrix} 1 & 0 \\ 0 & 1 \end{pmatrix},$$

$$e^0 \mapsto \gamma^0 := \begin{pmatrix} 0 & 1 \\ 1 & 0 \end{pmatrix}, \tag{11.12}$$

$$e^1 \mapsto \gamma^1 := \begin{pmatrix} 0 & -1 \\ 1 & 0 \end{pmatrix}, \tag{11.13}$$

$$e^0 e^1 \mapsto \gamma_5 := \begin{pmatrix} 1 & 0 \\ 0 & -1 \end{pmatrix}. \tag{11.14}$$

Note the numbering of the basis e^i now ranging from 0 to $n-1$, traditional whenever $r = 1$.

In four dimensions we have

$$C(3,1) \cong \mathbb{R}(4), \tag{11.15}$$

and in general, for even n there is an algebra isomorphism

$$\rho: C(r,s) \xrightarrow{\cong} \mathbb{R}(2^{n/2})$$

if and only if $r - s = 0$ modulo 8, e.g. in the case of $C(1,1)$, or $r - s = 2$ modulo 8, e.g. for $C(3,1)$. Note the asymmetry in r and s. The isomorphism ρ is called Majorana representation of the Clifford algebra. For any even n, there is an isomorphism from the complexified Clifford algebra onto $\mathbb{C}(2^{n/2})$:

$$\rho: C(r,s)^{\mathbb{C}} \to \mathbb{C}(2^{n/2}).$$

It is called the Dirac representation. The $\rho(e^i)$ are named γ-matrices:

$$\rho(e^i) =: \gamma^i. \tag{11.16}$$

We shall restrict ourselves to representations for even dimensions n, because only there do chiral spinors exist. For the general case we recommend the article by Coquereaux (1982).

11.2 Clifford groups

The Clifford group $Pin(r,s)$ is by definition the group generated by Clifford multiplication starting from the vectors $v \in C^1 = V$ of 'unit length',

$$g(v,v) = \pm 1. \tag{11.17}$$

$Pin(r,s)$ is a submanifold of the vector space $C(r,s)$ and therefore a Lie group. The special Clifford group $Spin(r,s)$ is defined as the subgroup of $Pin(r,s)$ consisting of even elements:

$$Spin(r,s) := Pin(r,s) \cap C_e. \tag{11.18}$$

As an example consider $r = 2$, $s = 0$. Any vector of unit length can be written

$$v = \cos\theta e^1 + \sin\theta e^2. \tag{11.19}$$

In order to generate the Clifford group we have to multiply two of them:

$$(\cos\theta e^1 + \sin\theta e^2)(\cos\theta' e^1 + \sin\theta' e^2)$$

$$= \cos(\theta - \theta')\mathbb{1} - \sin(\theta - \theta')e^1 e^2. \tag{11.20}$$

We keep on multiplying:

$$(\cos\theta e^1 + \sin\theta e^2)(\cos\varphi\mathbb{1} + \sin\varphi e^1 e^2)$$

$$= \cos(\theta + \varphi)e^1 + \sin(\theta + \varphi)e^2 \tag{11.21}$$

and finally

$$(\cos\varphi\mathbb{1} + \sin\varphi e^1 e^2)(\cos\varphi'\mathbb{1} + \sin\varphi' e^1 e^2)$$

$$= \cos(\varphi + \varphi')\mathbb{1} + \sin(\varphi + \varphi')e^1 e^2. \tag{11.22}$$

Therefore $Pin(2,0)$ is isomorphic to $O(2)$. It consists of two circles parametrized by θ and φ representing inversion-rotations and rotations, respectively. $Spin(2,0)$ is isomorphic to $SO(2)$, the second circle.

In the case $r = 1$, $s = 1$, we get similar results:

$$(\cosh\theta e^0 + \sinh\theta e^1)(\cosh\theta' e^0 + \sinh\theta' e^1)$$

$$= \cosh(\theta - \theta')\mathbb{1} - \sinh(\theta - \theta')e^0 e^1 \tag{11.23}$$

and

$$Spin(1,1)^e \cong SO(1,1)^e. \tag{11.24}$$

In low dimensions there are some more 'accidental' isomorphisms:

$$Spin(0,3) \cong SU(2), \tag{11.25}$$

$$Spin(1,3)^e \cong SL(2,\mathbb{C}), \tag{11.26}$$

which is often used to write spinors in $1 + 3$ dimensions as two component objects,

$$Spin(0,6) \cong SU(4), \tag{11.27}$$

the starting point of the twistor formalism. In higher dimensions the Clifford groups are not isomorphic to any classical matrix group.

We shall now construct a group homomorphism

$$\varphi: Pin(r,s) \to O(r,s).$$

As in the case of the three-dimensional rotation group (section 8.5) this is

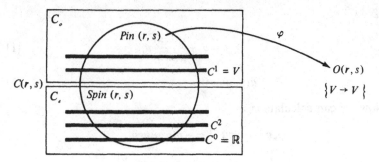

Fig. 11.1. A Clifford algebra and some subsets.

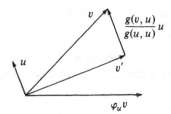

Fig. 11.2. Constructing the universal covering of $O(r, s)$.

achieved by a linear representation φ of $Pin(r,s)$ on V:

$$Pin(r,s) \times V \to V$$

$$(u, v) \mapsto \varphi_u v, \qquad (11.28)$$

$$\varphi_u v := uvu^{(-1)^{\text{degree }(u)}}.$$

A priori, $\varphi_u v$ is an element of $C(r, s)$. We have to show that it is indeed in $C^1 = V$. Note that all elements introduced so far, real numbers, vectors, algebra and group elements, are contained in one and the same set $C(r, s)$. This is often a source of confusion, which perhaps fig. 11.1 may help to avoid.

It is sufficient to show that $\varphi_u v$ belongs to V for any generator u of the Clifford group:

$$u \in C^1, \quad g(u, u) = \pm 1. \qquad (11.29)$$

Let us decompose v into its parts parallel to u and orthogonal to u as in fig. 11.2:

$$v = v' + \frac{g(v,u)}{g(u,u)} u, \tag{11.30}$$

$$g(v',u) = 0. \tag{11.31}$$

Now we can calculate $\varphi_u v$:

$$\varphi_u v = -uvu^{-1} = -g(u,u)uvu$$

$$= -g(u,u)uv'u - g(u,u)u\frac{g(v,u)}{g(u,u)}uu$$

$$= g(u,u)uuv' - \frac{g(v,u)}{g(u,u)}u$$

$$= v' - \frac{g(v,u)}{g(u,u)}u \quad \in V, \tag{11.32}$$

indeed a vector. Furthermore, $\varphi_u v$ is obtained from v by a reflection with respect to the plane orthogonal to u followed by an inversion. Therefore φ_u preserves angles, $\varphi_u \in O(r,s)$, and

$$\varphi: Pin(r,s) \rightarrow O(r,s)$$

is a group homomorphism. It is surjective, because any element of $O(r,s)$ can be written as a product of n reflections with respect to unit vectors. But φ is not injective, it is two-to-one,

$$\ker \varphi = \{\mathbb{1}, -\mathbb{1}\}. \tag{11.33}$$

Furthermore, $Spin(r,s)^e$ is simply connected for all $n \geqslant 3$. Hence in these cases it is the universal covering group of the proper Lorentz group $SO(r,s)^e$ and φ is the homomorphism of the second theorem in section 8.5.

11.3 Spinor representations

Recall that any representation of $SO(r,s)^e$ yields a representation of its universal covering group, just by composition with φ. A spinor or double-valued representation of $SO(r,s)^e$ is by definition a linear representation of $Spin(r,s)^e$ that cannot be obtained from a representation of $SO(r,s)^e$. Since the Clifford group is a subset of the Clifford algebra with the same product, the Majorana and Dirac representations ρ of the algebra carried by $\mathbb{R}^{2^{n/2}}$ and $\mathbb{C}^{2^{n/2}}$, respectively, (n even) are at the same time group representations. They are spinor representations and the vectors of $\mathbb{R}^{2^{n/2}}(\mathbb{C}^{2^{n/2}})$ are called Majorana (Dirac) spinors.

The Dirac representation, although irreducible as representation of the Clifford algebra, is reducible under $Spin(r,s)$. Indeed, $e^1 e^2 \cdots e^n \in C^n$ anticommutes with all elements of C^1. Therefore it commutes with all elements of C_r, in particular with all elements of $Spin(r,s)$ and can be used to reduce the Dirac spinors: We define

$$\gamma_5 := \rho(e^1 e^2 \cdots e^n) \tag{11.34}$$

which for $n \geqslant 6$ should not be confused with γ^5.

Since

$$\gamma_5^2 = (-1)^{(1/2)n(n-1)+s} \, \mathbb{1} \tag{11.35}$$

the following operators are projectors:

$$L := \tfrac{1}{2}(\mathbb{1} - \sqrt{(-1)^{(n(n-1)/2+s)}}\gamma_5), \tag{11.36}$$

$$R := \tfrac{1}{2}(\mathbb{1} + \sqrt{(-1)^{(n(n-1)/2+s)}}\gamma_5), \tag{11.37}$$

and a Dirac spinor $\psi \in \mathbb{C}^{2^{n/2}}$ splits into two $Spin(r,s)$-invariant components

$$\psi_L := L\psi, \tag{11.38}$$

$$\psi_R := R\psi, \tag{11.39}$$

called left- and right-handed Weyl (or chiral) spinors. In the case where $n(n-1)/2 + s$ is even, the above square root is real and the reduction is also possible for Majorana spinors yielding so-called Majorana–Weyl spinors. In conclusion, these exist if and only if $r - s = 0$ modulo 8, e.g. $r = s = 1$ where

$$L = \tfrac{1}{2}(\mathbb{1} - \gamma_5) = \begin{pmatrix} 0 & 0 \\ 0 & 1 \end{pmatrix}, \tag{11.40}$$

$$R = \tfrac{1}{2}(\mathbb{1} + \gamma_5) = \begin{pmatrix} 1 & 0 \\ 0 & 0 \end{pmatrix}. \tag{11.41}$$

We end this section with a remark on GL_n. Like $SO(n)$, the general linear group GL_n is not simply connected. However, its universal covering group has no linear representations other than GL_n representations (Cartan 1938). This is why physicists tried in vain for some time to define spinors in curved space using Einstein's gauge.

11.4 The Lie algebra of a Clifford group

To compute the Lie algebra $spin(r,s)$ of a Clifford group we use the one-parameter subgroups with elements

$$g_\theta := c(\theta)\mathbb{1} + s(\theta)e^i e^j, \tag{11.42}$$

where i and j are fixed. Depending on the signs of $g(e^i, e^i)$ and $g(e^j, e^j)$, $c(\theta)$ ($s(\theta)$) stands for $\cos\theta$ or $\cosh\theta$ ($\sin\theta$ or $\sinh\theta$) and the subgroup is $Spin(2,0) \cong Spin(0,2)$ or $Spin(1,1)$. A basis of the Lie algebra of $Pin(r,s)$ is then

$$\left.\frac{d}{d\theta} g_\theta\right|_{\theta=0} = e^i e^j, \quad i < j. \tag{11.43}$$

Therefore $spin(r,s)$ is just C^2, again contained in the Clifford algebra. Its dimension is $\frac{1}{2}n(n-1)$ and its bracket is the commutator of Clifford products. Its general element is

$$A = \tfrac{1}{4}\sum_{i,j}\Omega_{ij}e^i e^j, \quad \Omega_{ij} = -\Omega_{ji}. \tag{11.44}$$

We can use the group homomorphism φ to construct a Lie algebra isomorphism between $spin(r,s)$ and $so(r,s)$. Let us calculate the transformation of a vector $v \in C^1$,

$$v = \sum_k v_k e^k, \tag{11.45}$$

under a one-parameter subgroup:

$$\begin{aligned}\varphi_{g_\theta}v &= (c(\theta)\mathbb{1} + s(\theta)e^i e^j)v(c(\theta)\mathbb{1} - s(\theta)e^i e^j) \\ &= \sum_{l,k} v_l(\Lambda^{-1})^l{}_k e^k\end{aligned} \tag{11.46}$$

with

$$(\Lambda^l{}_k) := \begin{pmatrix} 1 & & & & & & \\ & \ddots & & & & & \\ & & c(2\theta) & \cdots & s(2\theta)\eta^{ii} & & \\ & & \vdots & & \vdots & & \\ & & -s(2\theta)\eta^{jj} & \cdots & c(2\theta) & & \\ & & & & & \ddots & \\ & & & & & & 1 \end{pmatrix} \in SO(r,s). \tag{11.47}$$

Λ is a 'rotation' in the i-j plane by an angle 2θ and

$$\left.\frac{d}{d\theta}\varphi_{g_\theta}\right|_{\theta=0} v = 2(\eta^{jj}v_j e^i - \eta^{ii}v_i e^j). \tag{11.48}$$

So we obtain the following Lie algebra isomorphism:

$$T_1 : spin(r,s) \to so(r,s)$$

$$\tfrac{1}{4}\sum_{i,j}\Omega_{ij}e^i e^j \mapsto \left(\sum_k \eta^{ik}\Omega_{kj}\right).$$

Finally, the Lie algebra representation $\tilde{\rho}$ on $\mathbb{C}^{2^{n/2}}$ corresponding to the group representation ρ is

$$spin(r, s) \times \mathbb{C}^{2^{n/2}} \to \mathbb{C}^{2^{n/2}}$$

$$(A, \psi) \mapsto \tilde{\rho}_A \psi,$$

$$\tilde{\rho}_A \psi = \tfrac{1}{4} \sum_{i,j} \Omega_{ij} \gamma^i \gamma^j \psi, \tag{11.49}$$

and we see that the Lie algebra representation coincides with the Clifford algebra representation:

$$\tilde{\rho}_A = \rho(A). \tag{11.50}$$

We end this section with an aside. We have seen that the Clifford algebra $C(r, s)$ contained a Lie group $Pin(r, s)$ with Lie algebra $spin(r, s) = C^2$ also contained in $C(r, s)$. The Clifford algebra contains a second nonabelian Lie group, namely itself with multiplication law $\circ$ defined by

$$(e, o) \circ (e', o') := (e + e' + oo', o + o'),$$

$$e, e' \in C_e, \qquad o, o' \in C_o. \tag{11.51}$$

Its neutral element is the origin $(0, 0)$ and the inverse is given by

$$(e, o)^{-1} = (-e + oo, -o). \tag{11.52}$$

Its Lie algebra is once more the entire Clifford algebra with commutator

$$[(e, o), (e', o')] = (oo' - o'o, 0). \tag{11.53}$$

This construction also works with a degenerate bilinear form g. For the Grassmann algebra ($g = 0$) with $n = \infty$ the Lie group we arrive at is the group of supersymmetries in one-dimensional spacetime (Langouche & Schücker 1985).

11.5 The Dirac operator on open subsets of $\mathbb{R}^n$

Let $\mathcal{U}$ be an open subset of $\mathbb{R}^n$, n even, endowed with a metric g. So we have a (differentiable) family of n-dimensional vector spaces $(T_x \mathcal{U})^*$ indexed by the points x in $\mathcal{U}$ together with a family of metrics $\overset{*}{g}_x$. We can choose an orthonormal frame, i.e. a family of orthonormal bases $e^i(x)$, $i = 1, 2, \ldots, n$. Mapping each element of $(T_x \mathcal{U})^*$ onto the n-tuple of its expansion coefficients with respect to the basis $e^i(x)$ one defines for all $x \in \mathcal{U}$ an isometry I_x from $(T_x \mathcal{U})^*$ onto $\mathbb{R}^n$ with metric η. Taking $V = \mathbb{R}^n$ in the above construction we obtain a Clifford algebra $C(r, s)$ and a Clifford group

Spin (r, s). A spinor ψ is defined to be a 0-form on $\mathscr{U}$ with values in the vector space $\mathbb{C}^{2^{n/2}}$ which carries the representation ρ of $C(r, s)$. Furthermore we introduce a spin connection ω, a 1-form on $\mathscr{U}$ with values in $spin(r, s)$, transforming under an element $g \in {}^{\#}Spin(r, s)^e$ as

$$\omega' = g\omega g^{-1} + g\,dg^{-1}. \tag{11.54}$$

Using the isomorphism $spin(r, s) \cong so(r, s)$ we obtain the transformation law under $\Lambda \in {}^{\#}SO(r, s)^e$ with $\Lambda = \varphi(\pm g)$:

$$\omega' = \Lambda\omega\Lambda^{-1} + \Lambda\,d\Lambda^{-1}. \tag{11.55}$$

As usual, the covariant derivative $D\psi$ is a 1-form with values in $\mathbb{C}^{2^{n/2}}$, given by

$$D\psi = d\psi + \rho(\omega)\psi. \tag{11.56}$$

The 1-form $D\psi$ can be decomposed with respect to the frame e^i:

$$D\psi = \sum_j (i_{e_j} D\psi) e^j, \tag{11.57}$$

where $e_1, e_2, \ldots, e_n$ is the frame of vector fields to which the e^i are dual. According to (11.16), the γ-matrices are

$$\gamma^i = \rho(I_x e^i(x)). \tag{11.58}$$

Obviously they are constant. Finally we define the Dirac operator $\rlap{/}{D}$:

$$\rlap{/}{D}\psi := \sum_j \gamma^j i_{e_j} D\psi. \tag{11.59}$$

Just as the covariant derivative, the Dirac operator $\rlap{/}{D}$ is covariant under ${}^{\#}Spin(r, s)^e$: Let g be an element of ${}^{\#}Spin(r, s)^e$ and $\Lambda = \varphi_g \in {}^{\#}SO(r, s)^e$, where the matrix Λ is defined by

$$\varphi_g e^k = \sum_j (\Lambda^{-1})^k{}_j e^j.$$

We transform ψ and e into

$$\psi' = \rho_g \psi, \tag{11.60}$$

$$e'^j = \sum_k \Lambda^j{}_k e^k. \tag{11.61}$$

Then

$$(\rlap{/}{D}\psi)' = \sum_j \gamma^j i_{e'_j} (D\psi)'$$

$$= \sum_{j,k} \gamma^j (\Lambda^{-1})^k{}_j i_{e_k} (\rho_g D\psi)$$

$$= \sum_{j,k} ((\Lambda^{-1})^k{}_j \gamma^j) \rho_g i_{e_k} D\psi$$

$$= \sum_k (\rho_g \gamma^k \rho_g - 1) \rho_g i_{e_k} D\psi$$

$$= \rho_g \sum_k \gamma^k i_{e_k} D\psi = \rho_g \not{D}\psi. \qquad (11.62)$$

The identity

$$\sum_j (\Lambda^{-1})^k{}_j \gamma^j = \rho_g \gamma^k \rho_g - 1 \qquad (11.63)$$

used above is obtained from (11.28).

For explicit calculations of the derivative $d\psi$ we need a holonomic frame dx^μ while the γ-matrices are defined in the orthonormal frame e^i. Both frames are related by the GL_n gauge transformation γ, sometimes called vielbein (cf. (5.85)):

$$e^i = \sum_\mu \gamma^i{}_\mu dx^\mu. \qquad (11.64)$$

Of course, it is not to be confused with the γ-matrices. Now the Dirac operator can be written:

$$\not{D}\psi = \sum_{i,\mu} (\gamma^{-1})^\mu{}_i \gamma^i \left(\frac{\partial}{\partial x^\mu} + \tfrac{1}{4} \sum_{a,b} \omega_{ab\mu} \gamma^a \gamma^b \right) \psi. \qquad (11.65)$$

In Minkowski space, $\mathcal{U} = \mathbb{R}^4$, $g = \eta$, the holonomic frame dx^μ of Cartesian coordinates is orthonormal: $\gamma = 1$. In the torsionless case, where $\omega = 0$, the Dirac operator reduces to

$$\not{D}\psi = \sum_\mu \gamma^\mu \frac{\partial}{\partial x^\mu} \psi, \qquad (11.66)$$

an expression which is covariant under the conventional (x-independent) Lorentz transformations acting also on the argument x.

In general, Dirac's operator is a linear, first-order partial differential operator with nonconstant coefficients $(\gamma^{-1})^\mu{}_i$, $\omega_{ab\mu}$:

$$\not{D} : \Lambda^0(\mathcal{U}, \mathbb{C}^{2^{n/2}}) \to \Lambda^0(\mathcal{U}, \mathbb{C}^{2^{n/2}})$$

$$\psi \mapsto \not{D}\psi.$$

Upon reduction of Dirac spinors to Weyl spinors,

$$\psi = \psi_L + \psi_R, \qquad (11.67)$$

the Dirac operator decomposes into two pieces

$$\not{D}_{LR} : \{\psi_L\} \to \{\psi_R\},$$

$$\not{D}_{RL} : \{\psi_R\} \to \{\psi_L\}.$$

For positive metric, $s = 0$, and vanishing torsion $\rlap{/}D_{LR}$ is the adjoint of $\rlap{/}D_{RL}$. Mathematicians usually call only one of the two pieces the Dirac operator.

If curvature and torsion vanish the square of Dirac's operator has a simple expression:

$$\rlap{/}D^2\psi = \Box \psi. \tag{11.68}$$

This was Dirac's starting point when he wrote down his operator 'as a square root of the d'Alembert operator'.

11.6 The Dirac action

We want to construct an action which upon variation of ψ yields the free field equation

$$\rlap{/}D\psi = 0. \tag{11.69}$$

To this end we need a nondegenerate sesquilinear form $(\,,\,)$ on $\mathbb{C}^{2^{n/2}}$,

$$(\psi, \chi) \in \mathbb{C}, \quad \psi, \chi \in \mathbb{C}^{2^{n/2}},$$

which is invariant under the group $Spin(r, s)^e$:

$$(\rho_g\psi, \rho_g\chi) = (\psi, \chi) \tag{11.70}$$

for all $g \in Spin(r, s)^e$.

For example, if we choose the Dirac representation such that all γ-matrices are unitary, which is always possible, we can set for $r = n, s = 0$

$$(\psi, \chi) := \psi^+ \chi. \tag{11.71}$$

This positive-definite form can be used to equip the space of spinors $\Lambda^0(\mathcal{U}, \mathbb{C}^{2^{n/2}})$ with a Hilbert space structure. For $r = 1, s = n - 1$ we may take

$$(\psi, \chi) := \psi^+ \gamma^0 \chi. \tag{11.72}$$

It is sufficient to verify the invariance for each generator of $Spin\,(r, s)^e$, which is easy.

We denote by $\bar\psi$ the adjoint of ψ with respect to the chosen sesquilinear form and define the Dirac action:

$$S_D := i \int *(\bar\psi \rlap{/}D\psi) \in \mathbb{C}. \tag{11.73}$$

It is invariant under $*Spin(r, s)^e$. This shows once more that the corresponding field equation (11.69) is covariant.

Note that in general the Dirac action is a complex number. Its real part is

$$\text{Re}\, S_{\text{D}} = \tfrac{1}{2}\left[\, \mathrm{i}\int *(\bar{\psi}\not{D}\psi) + \overline{\mathrm{i}\int *(\bar{\psi}\not{D}\psi)}\,\right]$$

$$= S_{\text{D}} + \tfrac{1}{2}\mathrm{i}\int * \sum_{a,b} \bar{\psi}\gamma^a\psi\, T^b{}_{ab}. \tag{11.74}$$

The torsion term comes from a partial integration with the derivative acting on the $(\gamma^{-1})^{\mu}{}_i$ in equation (11.65). More explicitly, the Dirac action reads in four dimensions

$$S_{\text{D}} = \frac{\mathrm{i}}{3!}\int \sum_{a,b,c,d} \bar{\psi}\gamma^a \mathrm{D}\psi \wedge e^b \wedge e^c \wedge e^d \varepsilon_{abcd} \tag{11.75}$$

and its real part

$$\text{Re}\, S_{\text{D}} = S_{\text{D}} + \frac{\mathrm{i}}{4}\int \sum_{a,b,c,d} \bar{\psi}\gamma^a\psi\, T^b \wedge e^c \wedge e^d \varepsilon_{abcd}. \tag{11.76}$$

A Rarita–Schwinger spinor is by definition a 1-form with values in $\mathbb{C}^{2^{n/2}}$. Its action in four dimensions is

$$S_{\text{RS}} := \frac{\mathrm{i}}{3!}\int \sum_{a,b,c,d} \bar{\psi} \wedge \gamma^a\gamma^b\gamma^c\, \mathrm{D}\psi \wedge e^d \varepsilon_{abcd}. \tag{11.77}$$

Supergravity is the Einstein–Cartan theory with one Majorana–Rarita–Schwinger spinor as matter. Note that for Majorana spinors the Dirac and Rarita–Schwinger Lagrangians are automatically real if we leave out the factor i in front, but they are also exact (a total divergence); e.g., in one-dimensional spacetime the Dirac Lagrangian of a Majorana spinor would be $\psi\,\mathrm{d}\psi$. The solution is to postulate that the $2^{n/2}$ components of the Majorana spinor are not real numbers but odd elements of some (infinite-dimensional) Grassmann algebra. They therefore anticommute in compliance with the Pauli principle and their Lagrangian is no longer trivial. In turn, the action has grown from a real number to an even element of a Grassmann algebra and likewise in supergravity, frame, spin connection, curvature, and torsion take values in this algebra.

11.7 Spin structures

If we want to define the Dirac operator on a manifold M we are immediately confronted with the problem that one essential ingredient of our construction in section 11.5, a frame, is only defined locally. When going from one open subset to another overlapping one, orthonormal frames are glued together by means of $O(r,s)$-transformations, which we have to lift now to

$Pin(r, s)$-transformations using the double-valued homomorphism φ^{-1}. In other words, the signs from $\ker\varphi = \pm 1$ must be patched together in a consistent manner. We have already seen an example where the matching of signs was not possible in general: the orientation of a manifold described locally by $\pm e^1 \wedge e^2 \wedge \cdots \wedge e^n$. There the patching, if it existed, was unique on a connected manifold. Here we have uniqueness if the manifold is simply connected. Any such patching is called a spin structure.

Before we give the definition of spin structures in its natural setting, that of fibre bundles, let us discuss some examples. Of course, every parallelizable manifold has a spin structure. A manifold that can be described by an atlas consisting of only two charts $(\mathcal{U}_1, \alpha_1)$, $(\mathcal{U}_2, \alpha_2)$ such that the overlap $\mathcal{U}_1 \cap \mathcal{U}_2$ is connected has a spin structure, because one has to glue only once. Examples are the spheres S^n. In four dimensions there is a lucky coincidence:

THEOREM (Geroch 1968)

Any four-dimensional spacetime manifold which is not compact possesses a spin structure if and only if it is parallelizable.

Let M be an n-dimensional manifold with metric g of signature $r - s$, $r + s = n$. The structure group GL_n of its bundle of frames $F(M)$ can always be reduced to $O(r, s)$ by means of the Gram–Schmidt orthonormalization procedure, yielding the bundle of orthonormal frames. Manifolds M where this group can be reduced further to the proper Lorentz group $SO(r, s)^e$ are called time and space orientable. For these we define a spin structure to be an extension of the bundle of (time and space) oriented orthonormal frames $F(M, g)$ with structure group $SO(r, s)^e$ to a principal bundle $S(M, g)$ with structure group $Spin(r, s)^e$, i.e. a bundle map $(f, \varphi, \mathrm{id}_M)$,

$$
\begin{array}{ccc}
S(M,g) & \xrightarrow{\;f\;} & F(M,g) \\
\Big\downarrow{\scriptstyle \pi_S} & & \Big\downarrow{\scriptstyle \pi} \\
M & \xrightarrow{\;\mathrm{id}_M\;} & M
\end{array}
$$

$$
Spin(r,s)^e \xrightarrow{\;\;\varphi\;\;} SO(r,s)^e,
$$

where φ is the group homomorphism constructed in section 11.2. This definition works not only in the case of even dimension n, to which most of our discussion was restricted. However we must exclude $n = 2$ because of the accidental isomorphisms $Spin(r, s) \cong SO(r, s)$ for $r + s = 2$.

By means of a given spin structure the Dirac operators defined locally in section 11.5 can be patched together on the entire manifold.

11.8 Kähler fermions

In this section we present a description of fermion fields which has the advantage that it works on an arbitrary Riemannian manifold M (Ivanenko & Landau 1928; Kähler 1962).

To begin with, we observe that the operator Δ (Laplacian or d'Alembert operator) acting on differential forms may be written as (cf. (3.46))

$$\Delta = (d - \delta)^2. \tag{11.78}$$

Therefore, $d - \delta$ is a 'square root' of Δ as is the Dirac operator in flat space, and one is led to consider the operator $d - \delta$ on inhomogeneous complex-valued differential forms Φ, i.e. on formal sums of complex-valued differential forms of various degrees:

$$\Phi = \sum_p \varphi^p, \quad \varphi^p \in \Lambda^p(M, \mathbb{C}). \tag{11.79}$$

In flat space there is a simple relation between $d - \delta$ and the Dirac operator. We explain it in the case of four dimensions. (The generalization to arbitrary even dimensions is obvious.) Let us introduce a 4×4 matrix Z of inhomogeneous differential forms:

$$Z := \mathbb{1} + \sum_{p=1}^{4} \sum_{\substack{\mu_1,\ldots,\mu_p = 1 \\ \mu_1 < \mu_2 < \cdots < \mu_p}}^{4} (-1)^{(1/2)p(p-1)}(\gamma_{\mu_1}\gamma_{\mu_2}\cdots\gamma_{\mu_p})^T$$

$$\times dx^{\mu_1} \wedge dx^{\mu_2} \wedge \cdots \wedge dx^{\mu_p}. \tag{11.80}$$

Here x^μ are Cartesian coordinates, and the index of the γ-matrices has been lowered by means of the metric. Each Φ can be expanded in terms of the differential forms Z_{ab} $(a, b = 1, \ldots, 4)$ with complex-valued coefficient functions $\psi^b{}_a$,

$$\Phi = \sum_{a,b=1}^{4} \psi^b{}_a Z_{ab}, \tag{11.81}$$

and one finds

$$(d - \delta)\Phi = \sum_{\mu=1}^{4} \sum_{a,b=1}^{4} (\gamma^\mu \partial_\mu \psi^b)_a Z_{ab}. \tag{11.82}$$

Here the lower index of ψ is considered as a spinor index. So we may interpret the 'Kähler field' Φ as describing four Dirac spinors $\psi^1, \ldots, \psi^4$.

For further details we must refer to the literature, e.g. (Graf 1978; Becher & Joos 1982; Benn & Tucker 1983).

Let us close with two remarks. First, there exists a natural (from the geometric point of view) lattice version of $d - \delta$ for the flat Euclidean metric, whereas this is not the case for the Dirac operator (Becher & Joos 1982). Secondly, $d - \delta$ is well defined on any Riemannian manifold. However, a relation to the Dirac operator like (11.82) holds only in flat space.

Problems

11.1 Construct an isomorphism between $Spin(0, 3)$ and $SU(2)$.

11.2 Calculate the square of γ_5.

11.3 Calculate the square of the Dirac operator in a flat, torsion-free space.

11.4 Show that (11.72) defines an invariant sesquilinear form.

11.5 Verify the second equation of (11.74).

12

An algebraic approach to anomalies

Anomalies are said to occur when symmetries of a classical theory are broken by quantum corrections. This breaking may be welcome as in the case of some approximate rigid symmetries. For example the axial $U(1)$-anomaly permits us to understand the π^0 decay into two photons. We shall be concerned only with the opposite case of 'sacred' symmetries necessary for consistency. The most prominent examples are gauge symmetries of quantized Yang–Mills theories in four-dimensional Minkowski space. Here the gauge symmetry ensures unitarity. Therefore one insists on vanishing anomalies which puts nontrivial constraints on the possible matter content, especially in the sector of chiral fermions. Continuous symmetries fall into two categories: the infinitesimal transformations, close to the identity of the symmetry group, and the global transformations. Anomalies of global transformations are more difficult to compute (Witten 1985) than infinitesimal anomalies, which we shall consider. Anomalies are defined in the context of quantum theory. Quantization of a field theory is beyond the scope of this text. However, there is a purely algebraic algorithm classifying infinitesimal anomalies, which is the subject of this chapter. Chapter 13 collects some explicit results of anomaly calculations via Feynman graphs.

12.1 Polynomials and Ward operators

At this point we have to formalize somewhat the notion of a Lagrangian and its invariance under a Lie algebra. Let φ^i, $i = 1, 2, \ldots, N$, 'the monomials', be real-valued differential forms on an open subset $\mathcal{U}$ of $\mathbb{R}^n$. We define the infinite-dimensional real vector space Pl of polynomials in the φ^i and their exterior and interior derivatives. The coefficients are from $\Lambda\mathcal{U}$, the product is the wedge product, and the inner derivatives are taken with respect to some given vector fields on $\mathcal{U}$. The vector space Pl is a graded vector space,

$$Pl = \bigoplus_{q=0}^{n} Pl_q, \qquad (12.1)$$

where q is the degree of the polynomial as differential form.

Let $\mathfrak{g}$ be a Lie algebra of dimension d, T_i, $i = 1, 2, \ldots, d$, a basis with structure constants $f_{kl}{}^i$ (cf. (4.78)). Let

$$A = \sum_{i=1}^{d} A^i T_i \in \Lambda^1(\mathcal{U}, \mathfrak{g}) \tag{12.2}$$

be a gauge potential. Take as monomials the d 1-forms A^i and the n 1-forms e^a of an orthonormal frame on $\mathcal{U}$. The components of the field strength

$$F = \sum_{i=1}^{d} F^i T_i \in \Lambda^2(\mathcal{U}, \mathfrak{g}) \tag{12.3}$$

are elements of Pl_2:

$$F^i = dA^i + \tfrac{1}{2} \sum_{k,l=1}^{d} f_{kl}{}^i A^k \wedge A^l \in Pl_2. \tag{12.4}$$

The Lagrangian

$$\mathcal{L} = \frac{1}{g^2} \operatorname{tr}(F \wedge *F) \tag{12.5}$$

is in Pl_n, where the trace is understood in some linear representation of $\mathfrak{g}$. The Hodge star needs some explanation: In order to be able to use its definition (3.16) in the context of polynomials, we have to write

$$F = \tfrac{1}{2} \sum_{a,b=1}^{n} (i_{e_b} i_{e_a} F) e^a \wedge e^b. \tag{12.6}$$

Now suppose that the φ^i carry a representation R of a – not necessarily finite-dimensional – Lie algebra $\mathfrak{E}$. In the above example $\mathfrak{E} = {}^{\mathcal{U}}\mathfrak{g}$, and $\Omega \in \mathfrak{E}$ is represented by

$$R_\Omega A^i = -d\Omega^i - \sum_{k,l=1}^{d} A^k \Omega^l f_{kl}{}^i, \tag{12.7}$$

$$R_\Omega e^a = 0, \tag{12.8}$$

where

$$\Omega = \sum_{i=1}^{d} \Omega^i T_i. \tag{12.9}$$

R induces a linear representation W on the vector space Pl: Let p be a polynomial from Pl, $E \in \mathfrak{E}$ an algebra element. Then we set:

$$W(E)p := [p(\varphi^i + f^i) - p(\varphi^i)]_{\text{lin}} \big|_{f^i = -R_E \varphi^i}. \tag{12.10}$$

The subscript lin means: Keep only terms linear in f^i. This procedure was already used to derive field equations from Lagrangians; see, e.g., (4.23), (5.29), (5.46). The mapping $W(E)$ is called the Ward operator and immediate consequences of its definition are:

$$W(E)\varphi^i = -R_E \varphi^i, \tag{12.11}$$

$$W(E)(p \wedge p') = (W(E)p) \wedge p' + p \wedge W(E)p', \tag{12.12}$$

$$dW(E)p = W(E)\,dp, \tag{12.13}$$

$$i_v W(E)p = W(E)i_v p, \quad v \in \text{vect}(\mathcal{U}). \tag{12.14}$$

Note that W is a linear representation even if R is not. In our example the A^i carry an affine representation. Still W is linear because the linearity of W is defined with respect to a different vector space structure, that of polynomials. As an example we calculate:

$$W(\Omega)F = [d(A+a) + \tfrac{1}{2}[A+a, A+a] - dA - \tfrac{1}{2}[A, A]]_{\text{lin}}\big|_{a = -R_\Omega A}$$

$$= (da + [A, a])\big|_{a = d\Omega + [A, \Omega]}$$

$$= d[A, \Omega] + [A, d\Omega] + [A, [A, \Omega]]$$

$$= [dA, \Omega] + [\tfrac{1}{2}[A, A], \Omega] = -[\Omega, F], \tag{12.15}$$

in agreement with (4.99). Note, however, the additional minus sign necessary in the representation W since 'its commutators are computed backwards'.

12.2 The Wess–Zumino consistency condition

A (classical) Lagrangian $\mathscr{L} \in Pl_n$ is called invariant under $\mathfrak{E}$ if $\int W(E)\mathscr{L} = 0$ for all $E \in \mathfrak{E}$. A Lagrangian is defined only up to 'surface terms', i.e. exact forms $d\chi$. Denoting the corresponding equivalence by

$$\mathscr{L} \sim \mathscr{L} + d\chi, \tag{12.16}$$

the invariance of $\mathscr{L}$ can be written

$$W(E)\mathscr{L} \sim 0. \tag{12.17}$$

The quantum theory is described perturbatively by an effective Lagrangian $\mathscr{L}_{\text{eff}}$ (cf. chapter 13). Because of loop integrations, the effective

Lagrangian contains products of fields at different spacetime points. It is therefore not 'local', i.e. $\mathscr{L}_{\text{eff}} \notin Pl$. In quantum theory Pl is enlarged to a bigger space B including also 'nonlocal' terms like $\mathscr{L}_{\text{eff}}$. The representation W of $\mathfrak{E}$ on Pl is extended to a representation $\bar{W}$ on B:

$$\bar{W}(E)|_{Pl} = W(E). \tag{12.18}$$

In general, some loop integrals in the effective Lagrangian diverge and have to be renormalized by adding appropriate local counterterms $p \in Pl_n$. Consequently, the effective Lagrangian is defined only up to exact forms and polynomials. We still denote this equivalence by $\sim$:

$$\mathscr{L}_{\text{eff}} \sim \mathscr{L}_{\text{eff}} + \mathrm{d}\chi + p. \tag{12.19}$$

For some theories any possible counterterm p suitable to renormalize divergent loops is not $\mathfrak{E}$-invariant:

$$W(E)p \neq \mathrm{d}\chi. \tag{12.20}$$

Such theories are said to be anomalous and their anomaly is defined by

$$\mathfrak{a}(E) := \bar{W}(E)\mathscr{L}_{\text{eff}}, \tag{12.21}$$

'the (anomalous) Ward identity'. Of course the anomaly is ambiguous,

$$\mathfrak{a}(E) \sim \mathfrak{a}(E) + \mathrm{d}\chi(E) + W(E)p, \tag{12.22}$$

which makes it difficult to decide whether two explicit results represent the same anomaly. Likewise it is difficult to see if an anomaly is trivial, $\mathfrak{a}(E) \sim 0$ for all $E \in \mathfrak{E}$, implying that there is an $\mathfrak{E}$-invariant way to renormalize the theory. Note that the anomaly itself is always local,

$$\mathfrak{a}(E) \in Pl_n, \tag{12.23}$$

because the bare, i.e. unrenormalized, effective Lagrangian is $\mathfrak{E}$-invariant. Using that $\bar{W}$ is a representation,

$$\bar{W}(E')\bar{W}(E) - \bar{W}(E)\bar{W}(E') = \bar{W}([E', E]), \tag{12.24}$$

we get immediately the Wess–Zumino consistency condition

$$W(E')\mathfrak{a}(E) - W(E)\mathfrak{a}(E') = \mathfrak{a}([E', E]) \text{ modulo exact forms} \tag{12.25}$$

(Wess & Zumino 1971), without any reference to the quantum extension B.

12.3 Stora's solutions

Although innocent looking, the Wess–Zumino consistency condition is quite restrictive. It is hard to find nontrivial solutions, the most prominent one being the Adler–Bardeen anomaly (Adler & Bardeen 1969; Bardeen 1969). Using cohomological methods, Dixon (unpublished) and Stora (1976) have constructed a class of nontrivial solutions for the case of gauge symmetries in four-dimensional Minkowski space. The Adler–Bardeen anomaly belongs to this class. Furthermore, Becchi, Rouet & Stora (1975; 1981) have shown that for any renormalizable gauge theory in four dimensions all solutions are obtained from Stora's algorithm.

We first give Stora's solutions in four dimensions. Let $\mathfrak{g}$ be any finite-dimensional Lie algebra and G the simply connected Lie group with Lie $G = \mathfrak{g}$. Let $I \in \mathscr{I}^3(G)$ be a symmetric invariant trilinear form on $\mathfrak{g}$. The invariance of I, (10.93), reads in infinitesimal form:

$$I([Z, A_1], A_2, A_3) + I(A_1, [Z, A_2], A_3) + I(A_1, A_2, [Z, A_3]) = 0 \quad (12.26)$$

for all $Z \in \mathfrak{g}$. A common example of such an invariant trilinear is the symmetrized trace over a linear representation $\tilde{\rho}$ of $\mathfrak{g}$:

$$I(A_1, A_2, A_3) = \frac{1}{3!} \sum_{\pi \in \mathscr{S}_3} \operatorname{tr} \{ \tilde{\rho}(A_{\pi(1)}) \tilde{\rho}(A_{\pi(2)}) \tilde{\rho}(A_{\pi(3)}) \}. \quad (12.27)$$

Let $\mathscr{U}$ be an open subset of $\mathbb{R}^4$ and consider the algebra of infinitesimal gauge transformations Ω,

$$\mathfrak{E} = {}^{\mathscr{U}}\mathfrak{g}. \quad (12.28)$$

If $A \in \Lambda^1(\mathscr{U}, \mathfrak{g})$ is a gauge potential, then

$$\mathfrak{a}(\Omega) := 3 \int_0^1 d\tau\, I(\Omega, F_\tau, F_\tau) - 6 \int_0^1 d\tau\, I(A, (\tau^2 - \tau)[\Omega, A], F_\tau) \in Pl_4 \quad (12.29)$$

solves the consistency condition, where

$$F_\tau := \tau\, dA + \tfrac{1}{2}\tau^2 [A, A] \quad (12.30)$$

is the field strength (10.100) with $\mathring{A} = 0$. Using the formula

$$\int_0^1 d\tau\, \tau^i = \frac{1}{i+1}, \quad i = 0, 1, 2, \dots, \quad (12.31)$$

the invariance and symmetry of I, we get

$$\mathfrak{a}(\Omega) = I(\Omega, dA, dA) + \tfrac{1}{4} I(\Omega, dA, [A, A]) + \tfrac{1}{2} I(\Omega, [dA, A], A). \quad (12.32)$$

For I given by (12.27), we recognize the Adler–Bardeen anomaly

$$\mathfrak{a}(\Omega) = \operatorname{tr} \{ \tilde{\rho}_\Omega d[\tilde{\rho}_A \wedge d\tilde{\rho}_A + \tfrac{1}{2}(\tilde{\rho}_A)^3] \}. \quad (12.33)$$

Since in this case all anomalies are of Stora's type, they are classified by the symmetric invariant trilinear forms. Consequently, a Lie algebra that does not admit any such form as, e.g., $su(2) \cong so(3)$ is 'safe', which means that all its gauge theories are anomaly-free for any matter content. Here matter means all monomials except connections and orthonormal frames. On the other hand, if the Lie algebra has an invariant form I, then of course any multiple cI, c a complex number, is also invariant. More generally, if $I_1, I_2, \ldots, I_r$ is a basis of $\mathcal{I}^3(G)$, then for a given matter content the anomaly is constructed from an invariant form

$$I = \sum_{i=1}^r c_i I_i, \quad c_i \in \mathbb{C}. \qquad (12.34)$$

The coefficients c_i have to be determined by other methods, discussed in chapter 13.

Note, however, that the anomaly given by a nonvanishing invariant is not necessarily nontrivial. This depends crucially on the matter content, from which also the local counterterms are constructed, and its transformation laws. A well-known example where a nontrivial gauge anomaly $\mathfrak{a}(\Omega)$ is rendered trivial by adding an extra field is the Wess–Zumino–Witten Lagrangian. The additional field g takes values in the Lie group G and transforms according to the representation given by left translation. The Wess–Zumino–Witten Lagrangian $\mathscr{L}_{\text{WZW}}(A, g)$ serves as counterterm to 'kill' the anomaly (Stora 1983; Mañes, Stora & Zumino 1985):

$$W(\Omega)\mathscr{L}_{\text{WZW}}(A, g) = \mathfrak{a}(\Omega). \qquad (12.35)$$

Another example is the famous anomaly cancellation noted by Green & Schwarz (1984), where nontrivial gauge anomalies in a ten-dimensional spacetime are rendered trivial by addition of a bosonic field with peculiar transformation properties.

We now generalize Stora's solutions to an arbitrary 'spacetime' manifold M of even dimension (Stora 1983; Zumino 1983),

$$\dim M = n = 2j - 2, \quad j = 2, 3, 4, \ldots, \qquad (12.36)$$

and arbitrary signature of the metric. We also admit gauge algebras of nontrivial bundles over M and include infinitesimal diffeomorphisms of M (Langouche, Schücker & Stora 1984). They can be lifted to the bundle by means of a fixed connection $\overset{\circ}{\mathscr{A}}$ as explained in section 9.14.

The total Lie algebra $\mathfrak{E}$ is described using a trivializing open covering $\{\mathscr{U}_r\}$ of M. On one such open subset $\mathscr{U}$ an element of $\mathfrak{E}$ is represented after pull back by a local section as a pair (Ω, v),

$$\Omega \in \Lambda^0(\mathcal{U}, \mathfrak{g}), \tag{12.37}$$

$$v \in \text{vect}(M), \tag{12.38}$$

and the bracket is defined by

$$[(\Omega', 0), (\Omega, 0)] = ([\Omega', \Omega], 0), \tag{12.39}$$

$$[(0, v'), (0, v)] = (-i_{v'} i_v \mathring{F}, [v', v]), \tag{12.40}$$

$$[(0, v), (\Omega, 0)] = (L_v \Omega + [i_v \mathring{A}, \Omega], 0). \tag{12.41}$$

Here the fixed connection $\mathscr{A}$ is represented on $\mathcal{U}$ by

$$\mathring{A} \in \Lambda^1(\mathcal{U}, \mathfrak{g}) \tag{12.42}$$

and its field strength by

$$\mathring{F} \in \Lambda^2(\mathcal{U}, \mathfrak{g}). \tag{12.43}$$

The role of $\mathscr{A}$ is to ensure that the commutators can be patched together on overlaps $\mathcal{U}_r \cap \mathcal{U}_s$. This is achieved by replacing the exterior derivative in the Lie derivative by a covariant exterior derivative $\mathring{D}$ with respect to $\mathring{A}$. Indeed:

$$[(0, v), (\Omega, 0)] = (\mathring{\mathscr{L}}_v \Omega, 0) \tag{12.44}$$

with

$$\mathring{\mathscr{L}}_v := i_v \mathring{D} + \mathring{D} i_v. \tag{12.45}$$

Note that different connections $\mathscr{A}$ on the same bundle yield isomorphic Lie algebras $\mathfrak{E}$. Note also that the gauge transformations $(\Omega, 0)$ form an ideal of $\mathfrak{E}$ while the vector fields $(0, v)$ in general do not form a subalgebra. However, if the bundle is trivial, $\mathring{A}$ is superfluous, and we put $\mathring{A} = 0$. Then $\mathfrak{E}$ is the semidirect product of $^M\mathfrak{g}$ and $\text{vect}(M)$.

The space of all connections on the bundle carries a representation of $\mathfrak{E}$ given locally by

$$R(\Omega, v)A = -d\Omega - [A, \Omega] + L_v A - di_v \mathring{A} - [A, i_v \mathring{A}]$$

$$= -D\Omega + i_v F + Di_v(A - \mathring{A}), \tag{12.46}$$

where the fixed auxiliary connection $\mathscr{A}$ does not transform under $\mathfrak{E}$:

$$R(\Omega, v)\mathring{A} = 0. \tag{12.47}$$

As before, R induces a linear representation W of $\mathfrak{E}$ on the space of polynomials Pl. Here, however, we must restrict ourselves to polynomials which are invariant under a change of the local trivialization transforming A and $\mathring{A}$ according to (9.99). Only these polynomials can be patched together and make sense on the bundle. Examples are (10.102) and (12.48) below.

Stora's solutions are global solutions of the Wess–Zumino consistency condition (12.25) for 𝔈. Their local expressions are

$$a(\Omega, v) = j \int_0^1 d\tau I(\Omega, F_\tau^{j-1})$$

$$- j(j-1) \int_0^1 d\tau I(A - \mathring{A}, (\tau^2 - \tau)[\Omega, A - \mathring{A}], F_\tau^{j-2})$$

$$- j(j-1) \int_0^1 d\tau I(A - \mathring{A}, (1-\tau)i_v\mathring{F}, F_\tau^{j-2}), \tag{12.48}$$

where $I \in \mathscr{I}^j(G)$ is a symmetric invariant j-linear form on $\mathfrak{g}$ and

$$F_\tau := d(\mathring{A} + \tau(A - \mathring{A})) + \tfrac{1}{2}[\mathring{A} + \tau(A - \mathring{A}), \mathring{A} + \tau(A - \mathring{A})] \tag{12.49}$$

the field strength (10.100) of the connection interpolating between $\mathring{A}$ and A. In principle one can of course show by brute force that (12.48) solves the consistency condition. A practicable proof uses cohomological methods and can be found in (Langouche, Schücker & Stora 1984) and (Schücker 1987). In this general case, however, it is not known whether there are solutions other than Stora's. Note that diffeomorphisms play a subordinate role in Stora's solutions in the sense that the infinitesimal diffeomorphism anomalies, the third term on the rhs of (12.48), vanish automatically for safe algebras.

12.4 Gravitational anomalies

Internal symmetries Ω as for instance in a Yang–Mills theory are characterized by an arbitrary finite-dimensional Lie algebra. They do not act on frames:

$$R(\Omega)e = 0, \tag{12.50}$$

or in the general setting of the last section,

$$R(\Omega, v)e = L_v e. \tag{12.51}$$

The gauge group of general relativity, on the other hand, is not represented on some abstract 'internal' space but on the tangent bundle. The underlying Lie algebra is the Lorentz algebra,

$$\mathfrak{g} = so(r, s), \quad r + s = n = 2j - 2, \tag{12.52}$$

and an element of the corresponding Lie algebra $\mathfrak{E}$ does act on orthonormal frames:

$$R(\Omega, v)e = \Omega e + \overset{\circ}{\mathscr{L}}_v e. \tag{12.53}$$

But Stora's solutions (12.48) are independent of any metric or orthonormal frame and therefore also valid for general relativity or any other theory of gravity invariant under the Lorentz gauge group. The Lorentz gauge anomalies are also called gravitational or Einstein anomalies. Due to the confusion of coordinate transformations with diffeomorphisms, the expression gravitational anomalies is sometimes used for diffeomorphism anomalies. We shall not follow this practice because diffeomorphism anomalies can also appear in pure Yang–Mills theories.

The Lie algebra $so(r, s)$ admits nonvanishing symmetric invariant k-forms if and only if k is even. This comes essentially from the fact that the trace over an odd power of an antisymmetric matrix vanishes. Consequently, there are no gravitational anomalies in four dimensions, $k = j = 3$.

In general, Stora (1983) has shown that two of his solutions constructed with two different connections A having identical field strength are equivalent. For gravitational anomalies this implies that they do not depend on torsion and may be calculated using the Riemannian connection.

Mixed (internal gauge and gravitational) anomalies may appear when an internal gauge theory is coupled to gravity. They are solutions of the Wess–Zumino consistency condition where the gauge group is the direct product of the internal and the Lorentz gauge group. For example, take a $U(1)$ Yang–Mills theory on a four-dimensional spacetime. Then mixed anomalies can be constructed from the following symmetric invariant trilinear form I on the total Lie algebra $u(1) \oplus so(1, 3)$:

$$I((A \oplus B)^3) := A \operatorname{tr} B^2, \quad A \in u(1), \quad B \in so(1, 3). \tag{12.54}$$

In higher-dimensional spacetimes there are of course more complicated mixed anomalies.

12.5 Lie algebra cohomology

In this last section we define the cohomology of Lie algebras and explain its connection with the Wess–Zumino consistency condition.

Let $\mathfrak{E}$ be a Lie algebra, finite-dimensional or not, and W a linear representation of $\mathfrak{E}$ on a vector space P. Consider the vector space

$$\Lambda(\mathfrak{E}, P) = \bigoplus_{l=0}^{\infty} \Lambda^l(\mathfrak{E}, P) \tag{12.55}$$

of alternating forms on $\mathfrak{E}$ with values in P. Define the coboundary operator

$$s: \Lambda^l(\mathfrak{E}, P) \to \Lambda^{l+1}(\mathfrak{E}, P)$$

$$Q \mapsto sQ$$

by

$$(sQ)(E_0, E_1, \ldots, E_l) := \sum_{i=0}^{l} (-1)^i W(E_i) Q(E_0, \ldots, \hat{E}_i, \ldots, E_l)$$

$$+ \sum_{\substack{i,j=0 \\ i<j}}^{l} (-1)^{i+j} Q([E_i, E_j], E_0, \ldots, \hat{E}_i, \ldots, \hat{E}_j, \ldots, E_l) \qquad (12.56)$$

where an argument with a circumflex is omitted. It follows that

$$s^2 = 0 \qquad (12.57)$$

justifying the name coboundary operator.

Just as for differential forms an alternating form Q on $\mathfrak{E}$ is called closed if

$$sQ = 0 \qquad (12.58)$$

and exact if

$$Q = sQ' \qquad (12.59)$$

for some form Q'. We define the subspaces of closed and exact l-forms,

$$Z^l(\mathfrak{E}, P) := \{Q \in \Lambda^l(\mathfrak{E}, P) | Q \text{ closed}\}, \qquad (12.60)$$

$$B^l(\mathfrak{E}, P) := \{Q \in \Lambda^l(\mathfrak{E}, P) | Q \text{ exact}\}, \qquad (12.61)$$

and

$$H^l(\mathfrak{E}, P) := Z^l(\mathfrak{E}, P)/B^l(\mathfrak{E}, P), \qquad (12.62)$$

the lth cohomology group of the Lie algebra $\mathfrak{E}$ with values in the representation space P.

If G is a (finite-dimensional) Lie group and $\mathfrak{E} = \mathfrak{g}$ its Lie algebra, then its Lie algebra cohomology is isomorphic to the de Rham cohomology of G restricted to left invariant differential forms. This isomorphism is proved using a formula valid for differential l-forms on any manifold M:

$$(d\psi)(v_0, v_1, \ldots, v_l) = \sum_{i=0}^{l} (-1)^i v_i(\psi(v_0, \ldots, \hat{v}_i, \ldots, v_l))$$

$$+ \sum_{\substack{i,j=0 \\ i<j}}^{l} (-1)^{i+j} \psi([v_i, v_j], v_0, \ldots, \hat{v}_i, \ldots, \hat{v}_j, \ldots, v_l),$$

$$v_i \in \text{vect}(M). \qquad (12.63)$$

Let us now consider $P = Pl_n$ and represent $\mathfrak{E}$ by the Ward operators

(12.10). In this situation the coboundary operator s is also called the **BRS** operator. Now, up to exact differential forms, the solutions of the Wess–Zumino consistency condition are in one-to-one correspondence with the elements of $H^1(\mathfrak{E}, Pl_n)$ where, as always, n is the dimension of spacetime: Let $\mathfrak{a} \in \Lambda^1(\mathfrak{E}, Pl_n)$ be a representative of a class in $H^1(\mathfrak{E}, Pl_n)$. Evaluating $\mathfrak{a}$ on a Lie algebra element $E \in \mathfrak{E}$ we obtain a polynomial $\mathfrak{a}(E) \in Pl_n$. For any polynomial

$$p \in \Lambda^0(\mathfrak{E}, Pl_n) = Pl_n \tag{12.64}$$

we have the equivalence

$$\mathfrak{a} \sim \mathfrak{a} + sp \tag{12.65}$$

or by application to E

$$\mathfrak{a}(E) \sim \mathfrak{a}(E) + W(E)p. \tag{12.66}$$

On the other hand, evaluating

$$0 = s\mathfrak{a} \in \Lambda^2(\mathfrak{E}, Pl_n) \quad \text{modulo exact forms} \tag{12.67}$$

on two Lie algebra elements E_0 and E_1 yields the Wess–Zumino consistency condition:

$$W(E_0)\mathfrak{a}(E_1) - W(E_1)\mathfrak{a}(E_0) - \mathfrak{a}([E_0, E_1]) = 0 \text{ modulo exact forms.} \tag{12.68}$$

Essential tools for constructing Stora's solutions are the Chern–Simons formula and the so-called (algebraic) Faddeev–Popov ghost z, which is the element of $\Lambda^1(\mathfrak{E}, Pl_0)$ defined by

$$z(\Omega) := \Omega \in Pl_0, \quad \Omega \in {}^{\mathscr{U}}\mathfrak{g}. \tag{12.69}$$

It resembles closely the Maurer–Cartan form, in particular

$$sz = -\tfrac{1}{2}[z, z]. \tag{12.70}$$

In this context the degree l in (12.55) is also called ghost number.

Problems

12.1 Prove the properties (12.12) and (12.13) of the Ward operator.
12.2 Show that the Adler–Bardeen anomaly (12.33) is of Stora's type.
12.3 Verify that (12.46) defines a representation of the Lie algebra $\mathfrak{E}$.
12.4 Calculate the square of the operator s defined in (12.56).

13
Anomalies from graphs

In the last chapter anomalies were treated algebraically as solutions to the Wess–Zumino consistency condition and as such had a precise mathematical meaning in the general context of a nontrivial bundle over an arbitrary even-dimensional spacetime. In this chapter we shall indicate how Feynman graphs are used to calculate infinitesimal gauge anomalies. Since quantization of fields on a curved manifold is still an open problem, we restrict ourselves to spacetimes that are vector spaces, $\mathbb{R}^n$. Although $\mathbb{R}^n$ admits only trivial bundles, quantizing a field theory in $\mathbb{R}^n$ is a subtle business, which lies beyond our scope. So we shall only present a sketchy outline and summarize the final results on the anomalies by exhibiting the corresponding symmetric invariant forms.

13.1 From the triangle to its trilinear form

Let spacetime be the four-dimensional flat Minkowski space, $\mathfrak{g}$ a finite-dimensional Lie algebra of an internal symmetry and $\tilde{\rho}$ a linear representation of $\mathfrak{g}$ on a complex vector space V. Consider the classical Lagrangian of a Yang–Mills theory coupled minimally to a multiplet of left-handed massless Dirac spin $\frac{1}{2}$ fermions $\psi \in \mathbb{C}^4 \otimes V$:

$$\mathscr{L} = \frac{1}{g^2} \operatorname{tr}(F \wedge *F) + i* \left[\bar{\psi} \left(\frac{\partial}{\partial x^\mu} + \tilde{\rho}(A_\mu) \right) \gamma^\mu \right.$$

$$\left. \times \tfrac{1}{2}(\mathbb{1} - i\gamma_5)\psi \right]. \tag{13.1}$$

After gauge fixing, the first term yields the propagator of the gauge bosons and their trilinear and quadrilinear self-couplings shown in fig. 13.1. The second term gives the fermion propagators and their coupling to the gauge bosons, fig. 13.2.

The effective Lagrangian contains in addition an infinite number of loop graphs, some of which are shown in fig. 13.3. Many of them diverge and have to be regularized. The theory is anomalous if all possible regularization procedures break the gauge symmetry $^{\mathbb{R}^4}\mathfrak{g}$ even after removal of the

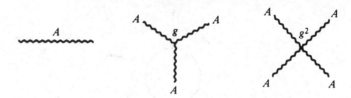

Fig. 13.1. Gauge boson propagator and couplings.

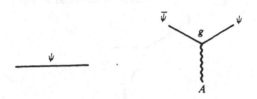

Fig. 13.2. Fermion propagator and coupling.

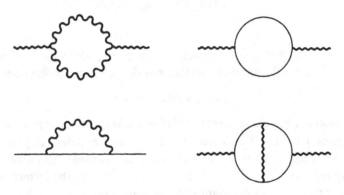

Fig. 13.3. Some loop graphs.

regularization. Clearly it is difficult to decide whether a theory really is anomalous. The following fact, however, considerably simplifies this task: Any loop graph containing at least one internal boson line can be regularized without breaking the gauge symmetry. This has two interesting consequences. First, only one-loop graphs with a fermion circulating on the internal line contribute to the anomaly and when calculating anomalies we can treat the bosons A classically. This means that we can drop the Yang–Mills term from the Lagrangian (13.1) making A an 'external', i.e. nonpropagating, background field. Secondly, graphs that contain external

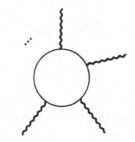

Fig. 13.4. Candidates for anomalous graphs.

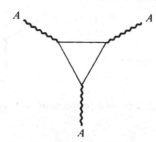

Fig. 13.5. The triangle graph.

fermion lines are never anomalous. Therefore, although fermions are responsible for the anomaly, it does not depend itself on fermions:

$$\mathfrak{a}(\Omega, A, \psi) = \mathfrak{a}(\Omega, A). \tag{13.2}$$

Anomalies appear in parity-violating theories in connection with the projector $\frac{1}{2}(1 - i\gamma_5)$. For instance, the two most popular gauge-invariant regularization schemes, Pauli–Villars and dimensional regularization, are easily seen to fail in the presence of chiral fermions, the former because of the absence of an invariant fermion mass term, the latter because

$$\gamma_5 = \frac{1}{4!} \sum_{\mu_i} \varepsilon_{\mu_0 \mu_1 \mu_2 \mu_3} \gamma^{\mu_0} \gamma^{\mu_1} \gamma^{\mu_2} \gamma^{\mu_3} \tag{13.3}$$

cannot be continued from 4 to complex dimensions in a straightforward way. Indeed, the anomaly will always come out proportional to the parity-violating ε-symbol.

The graphs we have to calculate are those with one closed chiral fermion line and a certain number of external gauge bosons attached to it (see fig. 13.4), the simplest one having two bosons with polarization vectors, say A_α and A_β. In four dimensions this graph does not contribute to the anomaly, essentially because there is no way to contract the four antisymmetric indices of the ε-symbol with the two polarization vectors A_α, A_β and the one derivative $\partial/\partial x^\mu$ from the variation (4.96) of the connection.

The next candidate is the triangle, fig. 13.5. It has three fermion propagators, diverges linearly, and must be regularized for instance by point splitting. We are interested in its gauge variation, which yields a finite 'local' polynomial of third order in the As. In the language of section 12.2, the triangle is an element of B. By applying the Ward operator $\bar{W}(\Omega)$ to it, we obtain the Adler–Bardeen anomaly (Adler & Bardeen 1969; Bardeen 1969), an element of Pl_4:

$$a(\Omega, A) \propto \varepsilon^{\mu\alpha\beta\gamma} \, \mathrm{tr} \left\{ \Omega \frac{\partial}{\partial x^\mu} \left[\tilde{\rho}(A_\alpha) \frac{\partial}{\partial x^\beta} \tilde{\rho}(A_\gamma) \right. \right.$$

$$\left. \left. + \tfrac{1}{2} \tilde{\rho}(A_\alpha) \tilde{\rho}(A_\beta) \tilde{\rho}(A_\gamma) \right] \right\}. \tag{13.4}$$

The proportionality $\propto$ means equality up to multiplication by an overall nonzero factor which depends on the coupling constant and in the following section also on the dimension and signature of spacetime but is the same for all representations $\tilde{\rho}$. The anomaly is additive and the trace just means that we sum all triangle anomalies with the different left-handed fermions circulating. (Right-handed fermions would give the same anomaly, but with opposite sign.) All we need to retain from the result (13.4) is the symmetric invariant from I (see (12.27)),

$$I(A^3) \propto \mathrm{tr} \, \tilde{\rho}_A^3, \quad A \in \mathfrak{g}, \tag{13.5}$$

from which the full expression of the anomaly can be reconstructed by Stora's algorithm (section 12.3). Note that in this calculation the symmetry of I comes from the boson symmetry in the three external lines of the triangle.

Now let us consider the candidates with more than three external bosons. Due to the additional fermion propagators and their minimal coupling, these graphs are less divergent than the triangle and can be invariantly regularized if the triangle anomaly vanishes (Adler 1970).

We close this section with some examples. Quantum electrodynamics is based on the Lie group $U(1)$. The trace in (13.5) reduces to

$$\sum_i q_i^3, \tag{13.6}$$

where q_i are the electric charges of the left-handed fermions. This sum always vanishes because for any left-handed particle with charge q there is also the left-handed antiparticle with opposite charge $-q$. This anomaly cancellation works of course in general for so-called vectorlike Yang–Mills theories where left-handed particles and left-handed antiparticles transform according to the same linear representation, because such a theory is parity conserving. In the parity-violating $SU(2) \times U(1)$ theory of electroweak interactions, we only have to worry about the $U(1)$ factor, since $SU(2)$ is a safe group. Here the anomaly is proportional to

$$\sum_i y_i^3, \tag{13.7}$$

where the y_is are the weak hypercharges of the left-handed particles. They are minus one for the electron, two for the positron, minus one for the neutrino. Therefore the anomaly can only be cancelled by introducing new particles, e.g. quarks. Our last example is the $SU(5)$ model of 'grand unification', which is anomaly-free if the chiral fermions transform according to the reducible representation $\bar{5} + 10$ because

$$\mathrm{tr}_{\bar{5}} A^3 = - \mathrm{tr}_{10} A^3, \quad A \in su(5). \tag{13.8}$$

13.2 Yang–Mills anomalies in even-dimensional spacetimes

Although Yang–Mills theories in more than four dimensions are not renormalizable, their anomalies can be calculated (Frampton & Kephart 1983). In fact, since the kinetic term $(1/g^2)\,\mathrm{tr}\,(F \wedge *F)$ of the gauge bosons characterizing Yang–Mills theories is irrelevant for anomalies, the results of this section apply to any gauge theory with minimally coupled chiral fermions. The calculations are similar to that in four dimensions. The first graph contributing to the anomaly is the fermion loop with j external gauge bosons, where as always the dimension of spacetime is $n = 2j - 2$. For left-handed Dirac fermions in a linear representation $\tilde{\rho}$ the anomaly is given by

$$I(A^j) \propto \mathrm{tr}\,\tilde{\rho}_A{}^j, \quad A \in \mathfrak{g}. \tag{13.9}$$

The same result holds for Rarita–Schwinger fermions, while for Kähler fermions the corresponding multilinear invariant form is 2^{j-1} times the above one. In all three cases the anomaly for Majorana fermions, whenever they exist, is one half of the anomaly of the corresponding complex fermions.

13.3 Gravitational anomalies

The gauge group of general relativity is based on the Lorentz group or on

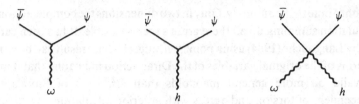

Fig. 13.6. The gravitational couplings to fermions.

$SO(n)$ in the Euclidean space where most graph calculations are performed. The theory is nonrenormalizable already in four dimensions. Still its anomalies can be calculated, but as in the case of Yang–Mills theories in more than four dimensions their physical relevance is not yet established. In gravity theories the couplings are not minimal. Indeed contrary to Yang–Mills theories every field couples, even if it transforms trivially under the gauge group, which here means that it has spin zero. The coupling of Dirac fermions to the gravitational background described by the frame and the spin connection ω is given by (11.74) or more explicitly by means of (11.65). It is common to most theories of gravity, not only Einstein's. Consequently, all results below are valid for these theories.

In order to separate propagators from couplings one has to use a weak field approximation for the metric. Expressed in terms of γ, (11.64), this approximation reads:

$$\det \gamma \ (\gamma^{-1})^\mu{}_i = \delta^\mu{}_i + h^\mu{}_i. \tag{13.10}$$

Now the gravitational background is described by the fields h and ω. Expanding the Dirac action (11.74) in terms of these fields we obtain in addition to the minimal coupling

$$\tfrac{1}{4} \sum_{\mu,a,b} \bar\psi \omega_{\mu ab} \gamma^\mu \gamma^a \gamma^b \psi \tag{13.11}$$

a derivative coupling

$$\sum_{\mu,i} \bar\psi h^\mu{}_i \gamma^i \frac{\partial}{\partial x^\mu} \psi \tag{13.12}$$

and the coupling

$$\tfrac{1}{4} \sum_{\mu,i,a,b} \bar\psi h^\mu{}_i \omega_{\mu ab} \gamma^i \gamma^a \gamma^b \psi \tag{13.13}$$

as represented in fig. 13.6.

Consequently, the degree of divergence of the fermion loop with k external bosons h or ω does not decrease with increasing k and *a priori* they all contribute to the anomaly. Only in two dimensions the complete anomaly calculation summing up all these graphs seems feasible. It has been carried out by Langouche (1984) using point splitting. His final result can be written in terms of the original variables of the Dirac action indicating that it might be valid on more general manifolds than $\mathbb{R}^2$. It is of Stora's type, independent of torsion and agrees with anterior calculations by Alvarez–Gaumé & Witten (1984). With two independent methods they have computed the gravitational anomalies of chiral Dirac and Rarita–Schwinger fermions and of the real self-dual tensor $\varphi \in \Lambda^{j-1}(\mathbb{R}^{2j-2})$,

$$* \varphi = \varphi, \tag{13.14}$$

in a space of arbitrary even dimension $n = 2j - 2$ and Minkowski signature. According to (3.18) this self-duality condition makes sense only if j is even. Recall that only in this case the gravitational anomalies of Stora's type (section 12.4) are nontrivial. In their first method Alvarez–Gaumé and Witten consider Feynman graphs under the following simplifications: Assuming zero torsion and working in the weak field approximation they calculate the graphs with j external bosons in the symmetric Lorentz gauge

$$\eta \gamma = (\eta \gamma)^T, \tag{13.15}$$

by reducing them to graphs of a scalar field in a constant electromagnetic background. Due to the gauge fixing, they obtain the anomaly here of course in a disguised form. In their second method they use the Atiyah–Singer index theorem in an auxiliary space of $n + 2$ dimensions. They get the same results as by the first method indicating that they are valid on more general manifolds. For more details we also recommend the article by Alvarez–Gaumé & Ginsparg (1985).

All results are of Stora's type given by the following j-linear forms for left-handed Dirac and Rarita–Schwinger fermions and the real self-dual tensor, respectively:

$$I_{1/2}(A^j) \propto \left[\prod_{i=1}^{j-1} \frac{\frac{1}{2}a_i}{\sinh \frac{1}{2}a_i} \right]_j, \tag{13.16}$$

$$I_{3/2}(A^j) \propto \left[\left(\prod_{i=1}^{j-1} \frac{\frac{1}{2}a_i}{\sinh \frac{1}{2}a_i} \right) \left(-1 + 2 \sum_{k=1}^{j-1} \cosh a_k \right) \right]_j, \tag{13.17}$$

$$I_{\varphi}(A^j) \propto \left[-\tfrac{1}{8} \prod_{i=1}^{j-1} \frac{a_i}{\tanh a_i} \right]_j. \tag{13.18}$$

Here $A \in so(1, n-1)$ is put into 'skew diagonal form',

$$\Lambda^{-1} A \Lambda = \begin{pmatrix} 0 & a_1 & & & & & \\ a_1 & 0 & & & & & \\ & & 0 & -a_2 & & & \\ & & a_2 & 0 & & & \\ & & & & \ddots & & \\ & & & & & 0 & -a_{j-1} \\ & & & & & a_{j-1} & 0 \end{pmatrix}, \Lambda \in SO(1, n-1), \quad (13.19)$$

and the index of the brackets means: After a Taylor expansion keep only the polynomials of degree j in the a_i. Again the anomaly of a Majorana fermion is half the anomaly of the corresponding complex fermion. The anomaly of a chiral Kähler fermion is the same as the anomaly of the self-dual tensor.

As an illustration we exhibit the results for ten dimensions:

$$I_{1/2}(A^6) \propto \frac{1}{10368}(\operatorname{tr} A^2)^3 + \frac{1}{4320}\operatorname{tr} A^2 \operatorname{tr} A^4 + \frac{1}{5670}\operatorname{tr} A^6, \qquad (13.20)$$

$$I_{3/2}(A^6) \propto \frac{-63}{10368}(\operatorname{tr} A^2)^3 + \frac{225}{4320}\operatorname{tr} A^2 \operatorname{tr} A^4 + \frac{-495}{5670}\operatorname{tr} A^6, \qquad (13.21)$$

$$I_{\varphi}(A^6) \propto \frac{64}{10368}(\operatorname{tr} A^2)^3 + \frac{-224}{4320}\operatorname{tr} A^2 \operatorname{tr} A^4 + \frac{496}{5670}\operatorname{tr} A^6, \qquad (13.22)$$

where the trace is in the fundamental representation. We recognize the celebrated anomaly cancellation in a theory with one right-handed Dirac spinor, one left-handed Rarita–Schwinger spinor, and one real self-dual tensor φ (Alvarez-Gaumé & Witten 1984).

Finally, mixed Yang–Mills and gravitational anomalies are constructed from invariant forms that are just products of the appropriate invariant forms from (13.9) and (13.16 to 13.18). In four dimensions, where there are no pure gravitational anomalies, let us consider the mixed anomaly of the $U(1)$ Yang–Mills theory of weak hypercharge. It comes from the triangle graph with one external $U(1)$ gauge boson A and two 'gravitons' ω (see fig. 13.7) and is given by

$$I((A \oplus \omega)^3) \propto \left(\sum_i y_i \right) \operatorname{tr} \omega^2. \qquad (13.23)$$

In the current $SU(2) \times U(1)$ model of electroweak interactions including quarks the weak hypercharges satisfy

$$\sum_i y_i = 0. \qquad (13.24)$$

13. Anomalies from graphs

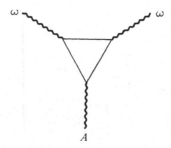

Fig. 13.7. The triangle with one gauge boson and two gravitons.

This has been taken as a hint towards 'grand unification' where $SU(2) \times U(1)$ is embedded together with the $SU(3)$ of strong interactions in a simple group implying a traceless hypercharge generator and hence (13.24). On the other hand, (13.24) is also equivalent to the vanishing of the mixed anomaly (Alvarez–Gaumé & Witten 1984).

Problems

13.1 Show that in four-dimensional spacetime the invariant (13.16) vanishes identically.

References

Adler, S. (1970). *Lectures on Elementary Particles and Quantum Field Theory*, Brandeis, ed. S. Deser *et al.* Cambridge Mass.: MIT Press.

Adler, S. & Bardeen, W.A. (1969). *Phys. Rev.* **182**, 1517.

Alvarez-Gaumé, L. & Ginsparg, P. (1985). *Ann. Phys.* **161**, 423.

Alvarez-Gaumé, L. & Witten, E. (1984). *Nucl. Phys.* **B234**, 269.

Arnowitt, R., Deser, S. & Misner, C.W. (1960a). *Phys. Rev.* **117**, 1595.

Arnowitt, R., Deser, S. & Misner, C.W. (1960b). *Phys. Rev.* **118**, 1100.

Arnowitt, R., Deser, S. & Misner, C.W. (1961). *Phys. Rev.* **122**, 997.

Bardeen, W.A. (1969). *Phys. Rev.* **184**, 1848.

Becchi, C., Rouet, A. & Stora, R. (1975). Lecture Notes, Erice, *Renormalization Theory*, ed. G. Velo, A.S. Wightman. Dordrecht: Reidel.

Becchi, C., Rouet, A. & Stora, R. (1981). *Field Theory, Quantization and Statistical Physics*, ed. E. Tirapegui. Dordrecht: Reidel.

Becher, P. & Joos, H. (1982). *Z. Phys.* **C15**, 343.

Belavin, A.A., Polyakov, A.M., Schwarz, A.S. & Tyupkin, Yu.S. (1975). *Phys. Lett.* **59B**, 85.

Benn, I.M. & Tucker, R.W. (1983). *Comm. Math. Phys.* **89**, 341.

Bogomolnyi, E. (1976). *Sov. J. Nucl. Phys.* **24**, 449.

Cartan, E. (1938). *Leçons sur la théorie des spineurs*. Paris: Hermann.

Coquereaux, R. (1982). *Phys. Lett.* **115B**, 389.

Deser, S. & Zumino, B. (1976). *Phys. Lett.* **62B**, 335.

Dirac, P.A.M. (1931). *Proc. Roy. Soc. London* **A133**, 60.

Dollard, J.D. & Friedman, C.N. (1979). *Product Integration*. Reading, Mass.: Addison-Wesley.

Eguchi, T. & Hanson, A.J. (1979). *Ann. Phys. (N.Y.)* **120**, 82.

Einstein, A. (1916). *Sitzungsberichte der Preußischen Akademie der Wissenschaften*, 1111.

Einstein, A. (1918). *Sitzungsberichte der Preußischen Akademie der Wissenschaften*, 448.

Einstein, A. & Infeld, L. (1940). *Ann. Math.* **41**, 455.

Einstein, A. & Infeld, L. (1949). *Canad. J. Math.* **1**, 209.

Einstein, A., Infeld, L. & Hoffmann, B. (1938). *Ann. Math.* **39**, 65.

Frampton, P.H. & Kephart, T.W. (1983). *Phys. Rev. Lett.* **50**, 1343.

Freed, D.S. & Uhlenbeck, K.K. (1984). *Instantons and Four-Manifolds*. New York: Springer.

Freedman, D., van Nieuwenhuizen, P. & Ferrara, S. (1976). *Phys. Rev.* **D13**, 3214.

Geroch, R. (1968). *J. Math. Phys.* **9**, 1739.

Graf, W. (1978). *Ann. Inst. H. Poincaré* **A29**, 85.

Green, M.B. & Schwarz, J.H. (1984). *Phys. Lett.* **149B**, 117.

Ivanenko, D. & Landau, L. (1928). *Z. Phys.* **48**, 340.

Kähler, E. (1962). *Rend. Mat. Ser. V*, **21**, 425.

Langouche, F. (1984). *Phys. Lett.* **148B**, 93.

Langouche, F. & Schücker, T. (1985). *Lett. Math. Phys.* **9**, 139.

Langouche, F., Schücker, T. & Stora, R. (1984). *Phys. Lett.* **145B**, 342.

Mañes, J., Stora, R. & Zumino, B. (1985). *Comm. Math. Phys.* **102**, 157.

Papapetrou, A. (1949). *Phil. Mag.* **40**, 37.

Polyakov, A. (1974). *JETP Lett.* **20**, 194.

Prasad, M. & Sommerfield, C. (1975). *Phys. Rev. Lett.* **35**, 760.

Rauch, H. *et al.* (1975). *Phys. Lett.* **54A**, 425.

Schlesinger, L. (1928). *Math. Ann.* **99**, 413.

Schoen, P. & Yau, S.T. (1979). *Phys. Rev. Lett.* **43**, 1457.

Schücker, T. (1987). *Comm. Math. Phys.* **109**, 167.

Stiefel, E. (1936). *Comm. Math. Helv.* **8**, 305.

Stora, R. (1976). Lecture Notes, Cargèse, *New Developments in Quantum Field Theory and Statistical Mechanics*, ed. M. Lévy & P. Mitter. New York: Plenum Press.

Stora, R. (1983). Lecture Notes, Cargèse, *Progress in Gauge Field Theory*, ed. G. 't Hooft *et al.* New York: Plenum Press.

't Hooft, G. (1974). *Nucl. Phys.* **B79**, 276.

Trautman, A. (1977). *Intern. J. Theor. Phys.* **16**, 561.

Uhlenbeck, K. (1979). *Bull. Amer. Math. Soc.* **1**, 579.

Wess, J. & Zumino, B. (1971). *Phys. Lett.* **37B**, 95.

Witten, E. (1985). *Comm. Math. Phys.* **100**, 197.

Yates, R.G. (1980). *Comm. Math. Phys.* **76**, 255.

Zumino, B. (1983). Lecture Notes, Les Houches, *Relativity, Groups and Topology II*, ed. B. De Witt & R. Stora. Amsterdam: North Holland.

Bibliography

Baum, H. (1981). *Spin-Strukturen und Dirac-Operatoren über pseudoriemannschen Mannigfaltigkeiten.* Leipzig: Teubner.

Bishop, R.L. & Crittenden, R.J. (1964). *Geometry of Manifolds.* New York: Academic Press.

Bourbaki, N. (1958). *Algèbre.* Paris: Hermann.

Chern, S.S. (1968). *Complex Manifolds without Potential Theory.* New York: Springer.

Chevalley, C. & Eilenberg, S. (1948). Cohomology theory of Lie groups and Lie algebras. *Trans. Amer. Math. Soc.* **63**, 85.

Choquet-Bruhat, Y., De Witt-Morette, C. & Dillard-Bleick, M. (1982). *Analysis, Manifolds and Physics.* Amsterdam: North Holland.

Daniel, M. & Viallet, C.M. (1980). The geometrical setting of gauge theories of the Yang–Mills type. *Rev. Mod. Phys.* **52**, 175.

Eguchi, T., Gilkey, P. & Hanson, A. (1980). Gravitation, gauge theories and differential geometry. *Phys. Rep.* **66**, 213.

Gilmore, R. (1974). *Lie Groups, Lie Algebras and Some of Their Applications.* New York: Wiley.

Kobayashi, S. & Nomizu, K. (1963). *Foundations of Differential Geometry*, vol. 1. New York: Wiley.

Kobayashi, S. & Nomizu, K. (1969). *Foundations of Differential Geometry*, vol. 2. New York: Wiley.

Steenrod, N. (1951). *The Topology of Fibre Bundles.* Princeton, N.J.: Princeton University Press.

Straumann, N. (1981). *Allgemeine Relativitätstheorie und relativistische Astrophysik.* Berlin: Springer.

Thirring, W. (1978). *Lehrbuch der mathematischen Physik.* Wien: Springer.

Weinberg, S. (1972). *Gravitation and Cosmology: Principles and Applications of the General Theory of Relativity.* New York: Wiley.

Notation

$\cong$	isomorphic	
$\bar{z}$	complex conjugate of $z \in \mathbb{C}$	
A^T	transpose of the matrix A	
A^+	hermitian conjugate of the complex matrix A	
$\mathbb{1}$	unit matrix, neutral element of a Clifford algebra	
$\otimes$	tensor product	
$\oplus$	direct sum	
ker	kernel of a homomorphism	
tr	trace of a square matrix	
$\mathscr{S}_n$	permutation group of n objects	
sig π	signature of $\pi \in \mathscr{S}_n$	
$\varepsilon_{i_1 \cdots i_n}$	ε-symbol	7
$\delta^i{}_j, \delta_{ij}$	Kronecker symbol	
$\delta(x)$	Dirac distribution	
(a, b)	open interval $\{x \in \mathbb{R} \mid a < x < b\}$	
$[a, b]$	closed interval $\{x \in \mathbb{R} \mid a \leqslant x \leqslant b\}$	
id_M	identity map of M onto itself	
$f\vert_{\mathscr{U}}$	restriction of the map f to $\mathscr{U}$	
V^*	dual space of the vector space V	1
$GL(V)$	group of automorphisms of the vector space V	
GL_n	group of real invertible $n \times n$ matrices	1
$GL(n, \mathbb{C})$	group of complex invertible $n \times n$ matrices	
$SL(n, \mathbb{C})$	group of complex $n \times n$ matrices with unit determinant	
$O(r, s)$	pseudoorthogonal group	34
$SO(r, s)$	special pseudoorthogonal group	36, 117
$U(n)$	group of unitary $n \times n$ matrices	
$SU(n)$	group of unitary $n \times n$ matrices with unit determinant	117

$Pin(r,s)$	Clifford group	187
$Spin(r,s)$	special Clifford group	187
$gl(V)$	Lie algebra of endomorphisms of the vector space V	124
$gl_n, so(n), \ldots$	Lie algebra of $GL_n, SO(n), \ldots$	121
$\mathbb{R}(n)$	associative algebra of real $n \times n$ matrices	
$\mathbb{C}(n)$	associative algebra of complex $n \times n$ matrices	
$\Lambda^p V$	space of alternating p-forms on the vector space V	2
$\Lambda^p(V,W)$	space of alternating p-forms on the vector space V with values in the vector space W	5
$\wedge$	wedge product	2
$[\varphi, \psi]$	bracket of Lie algebra valued forms	5
F^*	pull back	6, 15
$T_x M$	tangent space	100
TM	tangent bundle	101
$\dot{Q}$	tangent vector of a curve	102
$T_x F$	tangent mapping	104
$\dfrac{\partial}{\partial x^i}$	holonomic frame of vector fields	10, 100
dx^i	holonomic frame of 1-forms	14
$\Lambda^p M$	space of differential p-forms	105
$\Lambda^p(M,W)$	space of differential p-forms with values in the vector space W	105
i_v	inner derivative	4
d	exterior derivative	16
L_v	Lie derivative	84, 89
D	exterior covariant derivative	54, 147
δ	coderivative	40
$\not{D}$	Dirac operator	193
∂M	boundary of M	108
g	metric	33
$\overset{*}{g}$	induced metric on the dual space	34
g_{ij}	matrix of the metric g	33
g^{ij}	matrix of the metric $\overset{*}{g}$	34
g	$\det(g_{ij})$	40
η	$\mathrm{diag}(1,\ldots,1,-1,\ldots,-1)$	33
J	canonical isomorphism $V \to V^*$	35
$*$	Hodge star	33

$^M G$	gauge group	59
$^M \mathfrak{g}$	gauge algebra	59
$\mathrm{Diff}(M)$	group of diffeomorphisms	97
$\mathrm{vect}\,(M)$	Lie algebra of vector fields	102
$[v, w]$	Lie bracket	85, 102
$C^\infty(M)$	algebra of smooth functions	96
e	neutral element of a group	
G^e	connected component of e	116
$\rho_g = \rho(g)$	group representation	118
bit_x	orbit mapping	119
L_g	left translation	120
Aut_g	inner automorphism	120
R_g	right translation	120
$\tilde{R}_g$	right action on a principal bundle	135
$\mathfrak{g} = \mathrm{Lie}\,(G)$	Lie algebra	121
$\tilde{\rho}_A = \tilde{\rho}(A)$	Lie algebra representation	123
$\mathrm{Ad}_g, \mathrm{ad}_A$	adjoint representations	125
σ_a	Pauli matrices	127
τ_a	$-\frac{1}{2}i\sigma_a$	127
$F(M)$	frame bundle	139
λA	fundamental vector field	144
$C(r, s)$	Clifford algebra	185
γ^i	γ-matrices	187
γ_5	$\gamma^1 \gamma^2 \cdots \gamma^n$	191
$\mathscr{I}^j(G)$	space of symmetric invariant j-forms on $\mathfrak{g}$	179
$\propto$	proportionality	215

Index

Printed in the United States
By Bookmasters